# ROCKS, ICE & WATER

## The Geology of Waterton-Glacier Park

David D. Alt. Ph.D. and Donald W. Hyndman, Ph.D.
*Department of Geology, University of Montana*

*Published in cooperation with the*
*Glacier Natural History Association, Inc.*

MOUNTAIN PRESS PUBLISHING COMPANY
MISSOULA, MONTANA

Library of Congress Catalog Card Number 73-78910
ISBN 0-87842-041-X

*Cover Photograph by M.E. Lacy, Whitefish*
*Iceberg Lake in Glacier National Park*

Mountain Press Publishing Company
287 West Front Street
Missoula, Montana 59801

# Foreword

Glacier National Park in northwestern Montana and Waterton Lakes National Park in Alberta display a rugged beauty that provides a living memorial to international cooperation and goodwill.

These parks, comprising 1,800 square miles of the northern Rocky Mountains, exemplify the meaning of "natural park." Maintained here for the appreciation and understanding of mankind are picturesque peaks and valleys sculptured by glaciers long ago, nearly 50 living glaciers nestled in high natural ampitheaters, more than 200 lakes enlivened by clear streams, lush forest, a profusion of wildflowers, and an abundance of wildlife. All these things are interacting together in a delicate, dynamic community. This is truly a "living natural history museum."

Basic to this grandeur and interactions of living things is the story of the land. . .the earth itself. . .how it was formed and how it is changing.

Use this booklet as your guide to these parks. The authors, David Alt and Donald Hyndman, have skillfully and colorfully put together in nontechnical language a story that can deepen your understanding and appreciation of these two great parks. . .and of your own environment.

Ronald H. Walker
Director, National Park Service

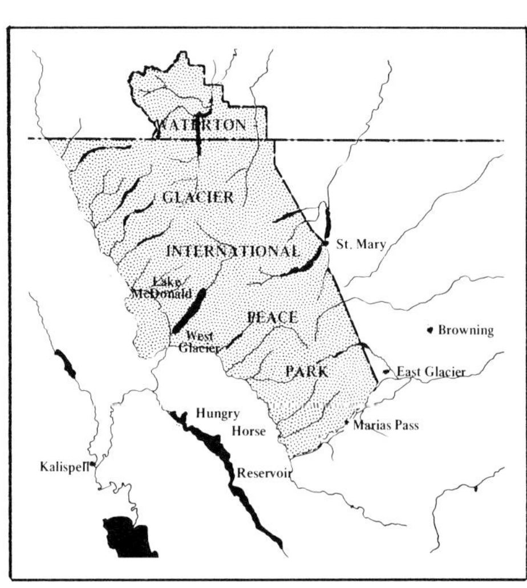


*Waterton-Glacier Park as it looks from space. A photograph taken in infra-red light at an elevation of about 500 miles by the ERTS-1 satellite. Photograph courtesy of The National Aeronautics and Space Administration.*

0 10 miles

# Preface

*The geology of Waterton-Glacier International Peace Park is the basis for its spectacular scenery. Geologic processes, both past and present, are fundamental to the ecological systems operating in the Park.*

*We have emphasized the geology most easily accessible from the park roads because most visitors rely on their cars for transportation. But hikers should find the book equally useful away from the roads because the geology of Waterton-Glacier Park is all cut from the same piece of cloth. Except for place names, this book would have turned out almost the same had the roads been built along entirely different routes.*

*Information in this book was derived from geologic literature supplemented by our own observations made over a period of several years. Studying and writing about Waterton-Glacier Park is a pleasant task made considerably more so by the assistance and cooperation of Chief Naturalists Edwin Rothfuss and Duane Barrus and their staffs in Glacier and Waterton Parks, respectively. Numerous professional colleagues have also contributed suggestions and comments, especially G.W. Crosby whose observations on the Lewis overthrust fault were most helpful.*

*Winter is never more than a few weeks away in Waterton-Glacier Park. Jim Mohler photo.*

# Table of Contents

# Introduction

These are splendid mountains.

Scenery has its foundation in geology and magnificent scenery is usually built on spectacular geology. The geology of Waterton-Glacier Park fully matches the promise of its scenery. Understanding the geology adds a dimension of meaning and excitement to the beauty of the park.

Here, well exposed, are some of the most ancient well-preserved sedimentary rocks known anywhere in the world. Some of them must have been deposited well over one billion years ago. Delicate sedimentary features and fossils of primitive plants are perfectly preserved, opening for us one of the few windows offering a clear view into the remote geologic past. In most other similarly ancient sedimentary rocks all traces of delicate structures have been obliterated by recrystallization clouding the view through the window.

Thousands of feet of ancient mudstones, limestones and sandstones contain evidence of sedimentary environments that were alternately wet and dry and locally scummy with growths of primitive blue-green algae. There are no traces of animals, a surprising discovery because the oldest known animal fossils are complex creatures that must have had ancestors living far back in Precambrian time when our rocks were laid down.

Quite possibly, we see in these rocks a record of deposition in a time when the earth's atmosphere and climate were different from what they have since become. So far, geologists can only speculate on what conditions may have been like when the Precambrian rocks of Waterton-Glacier Park were deposited. Certainly they were quite unlike any situation that we now know.

These Precambrian sediments remained essentially undisturbed for the better part of a billion years. At times, the region was shallowly flooded by invasions of the sea and more layers of sedimentary rock were deposited. Then, about 70 million years ago, the sea retreated for the last time as crustal upheavals raised the Rocky Mountain region several thousand feet and the accumulated sedimentary rocks in the region of Waterton-Glacier Park began to slide off the uplift. Great slabs of sedimentary rock slid eastward, moving very slowly over a period of millions of years, and piled on each other to form the eastern front of the Rockies. Mountains in the Park are carved by erosion from a single great slab of ancient sedimentary rock that slid so far east that it rode over soft deposits of mud and sand laid down during the last invasion by the sea.

For many millions of years after formation of the Rocky Mountains began, the region had a dry climate. Lack of stream drainage caused the big valleys to fill with thousands of feet of mud, sand and gravel washed out of the nearby mountains. Flashfloods sweeping over the countryside after occasional rains graded vast surfaces that stretched away unbroken across the high plains.

Approximately 3 million years ago the climate changed and the first of at least four major ice ages of the Pleistocene Epoch began. Streams began the work of excavating the accumulated sediment from the big valleys and carving the level surface of the high plains into rolling hills. Large glaciers gouged out the mountain valleys and carved their peaks, creating the fiercely jagged landscape we see today. Continental ice caps spread over much of Canada, extending southwestward to the Park. Such conditions existed as recently as 10,000 years ago.

Now the glaciers have almost entirely melted and other processes have begun the work of reshaping the landscape. Soil forming on rock slopes left by the ice is beginning to creep slowly down them, gradually transforming ragged cliffs into smoothly-rounded hillslopes. Landslides, rock falls, and avalanches are also leaving their marks as they move material downslope. Clear streams running thin along the floors of valleys once brimful of noisily grinding ice are carrying away the eroded sediment and carving their own delicate sculpture into the valley floor.

Past events and present processes have clearly marked the landscape of Waterton-Glacier Park. The moving rocks, ice, and water have left their signatures in the scenery of the Park.

# Precambrian Sedimentary Rocks

Erosion has carved the mountains of Waterton-Glacier Park from some of the oldest well-preserved sedimentary rocks known. They were originally deposited as layers of sand, silt, and mud during a span of time that must have begun well over one billion years ago and continued until about 600 million years ago. Despite their antiquity, these rocks are so perfectly preserved that they look almost as fresh as layers of sand and mud deposited yesterday, a marvelous glimpse into the remote geologic past. Geologists call them the "Belt" sedimentary rocks because they were first closely studied in the Belt Mountains of west-central Montana.

Although the "Belt" sedimentary rocks are beautifully preserved, the details of their origin are obscure. Most of the formations contain abundant evidence that the original sediments were deposited in an environment that was alternately wet and dry. But that might have been a delta, a tidal flat, a desert plain, or perhaps some other environment. Most likely some kind of situation that has no exact parallel in the modern world. Certainly the world was a very different place when the Belt rocks were deposited; there were no land plants and the atmosphere may not have been fit for animals to breathe. It is difficult for us to imagine the world as it was then and perfectly possible that we have not yet been able to reconstruct in our minds the environment in which the Belt rocks were formed.

### *Fossils*

Living things must have been at a rudimentary stage of development when the Belt rocks were deposited. Plant fossils are abundant in some layers, invariably remains of blue-green algae, one of the most primitive kinds of plants. Any animals that may have lived in those early days left no record of themselves in these rocks, at least none that anyone has been able to recognize.

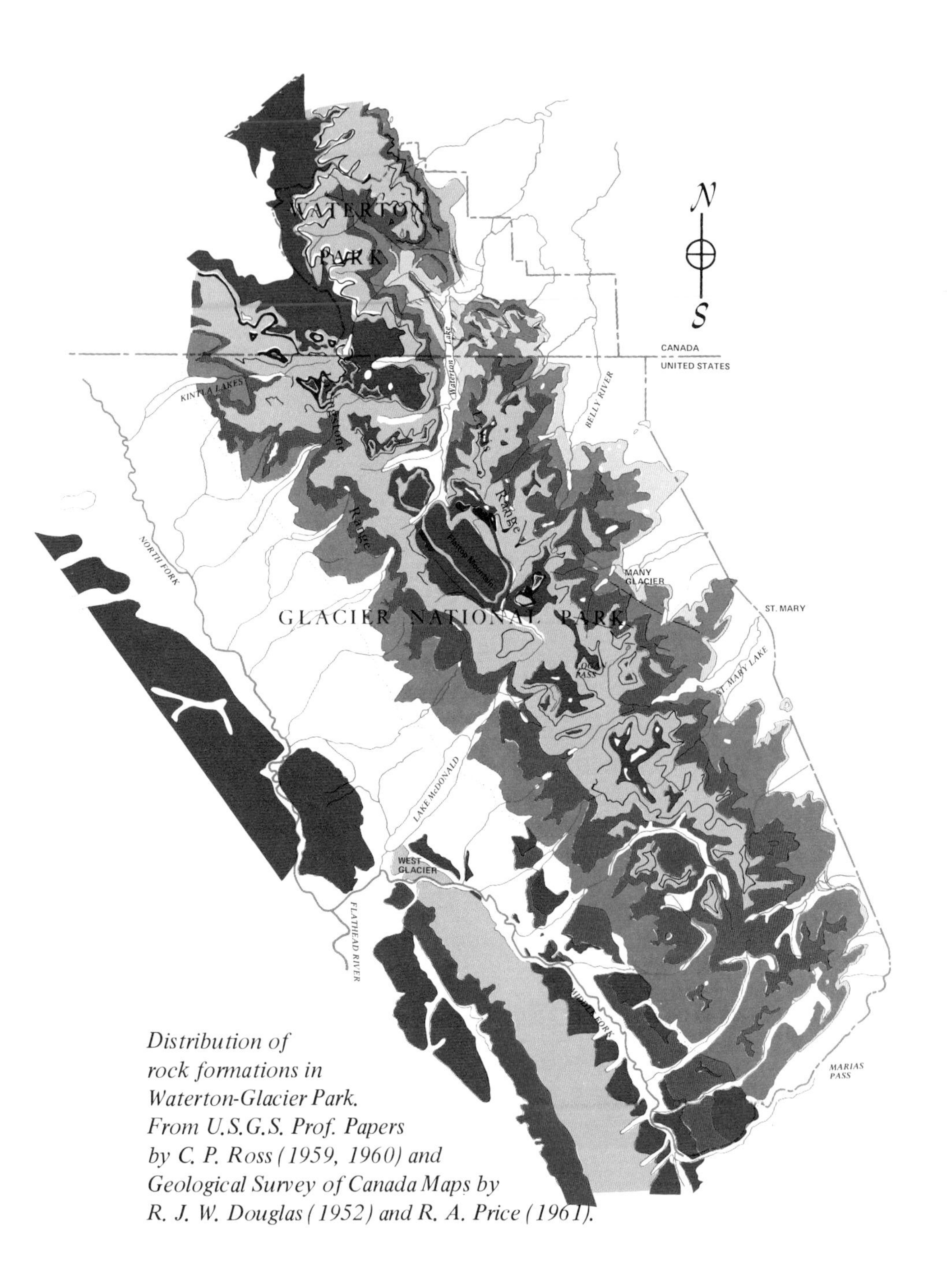


*Distribution of rock formations in Waterton-Glacier Park. From U.S.G.S. Prof. Papers by C. P. Ross (1959, 1960) and Geological Survey of Canada Maps by R. J. W. Douglas (1952) and R. A. Price (1961).*

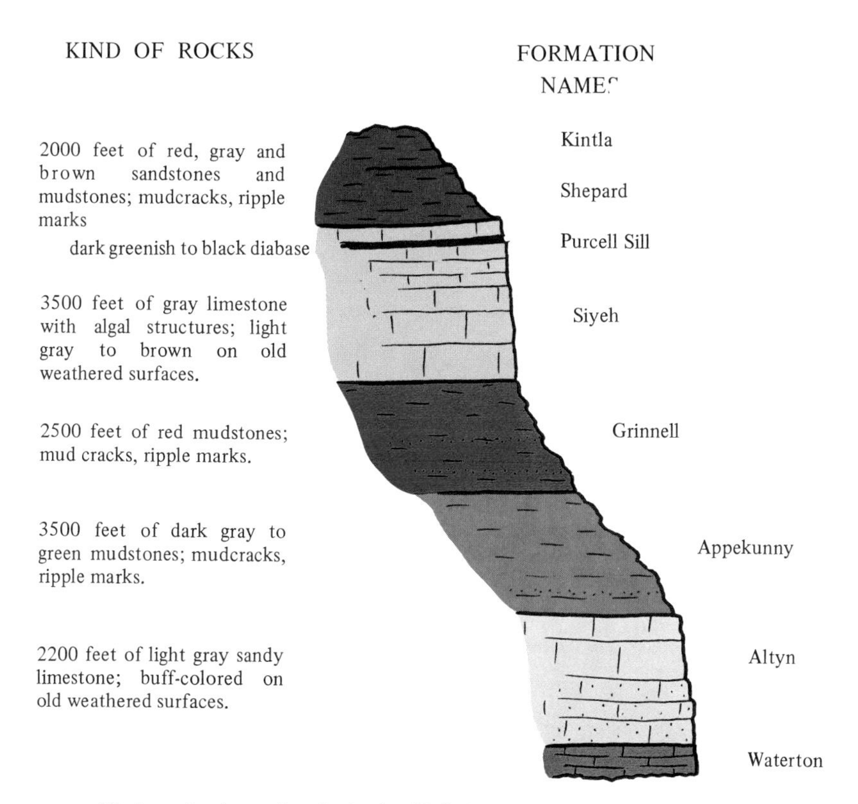


*Kinds and colors of rocks in the "Belt" formations of Waterton-Glacier Park.*

Absence of animal fossils is especially intriguing because it seems likely that animals of some kind must have lived when the Belt sedimentary rocks were laid down. The oldest known animal fossils are remains of highly-evolved creatures found in rocks deposited about 600 million years ago. Ancestors of these animals must have lived much earlier, perhaps they were soft-bodied animals not capable of forming fossils and leading a life-style that left no tracks on a muddy bottom. Or possibly there are fossils still waiting to be found somewhere in the many thousands of layers of Belt sedimentary rocks. Such a discovery would certainly be an outstanding contribution to our knowledge of the earliest forms of life on earth.

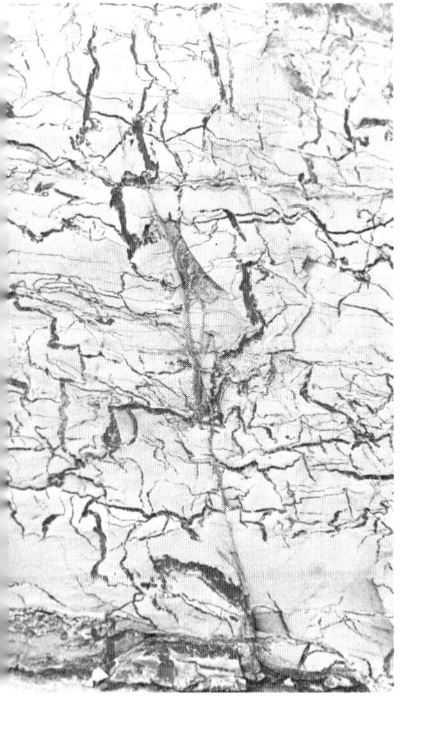

*Some of the many forms of fossil blue-green algae in the Siyeh Limestone.*

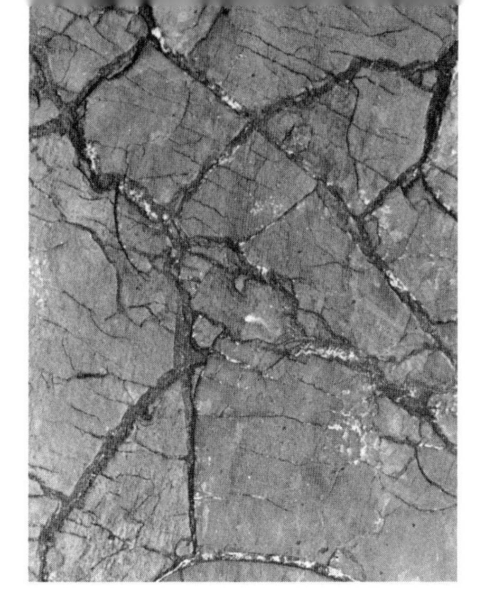

*Suncracked mud in Belt sedimentary rock. Absence of animals to disturb the mud during deposition may explain why such perfect preservation of delicate sedimentary structures is typical of these ancient rocks.*

Fossil blue-green algae occur in a bewildering variety of forms. Most often they are simply very thin layers in the rock, frequently they have the form of crumpled sheets that appear as irregular lines on a broken surface, and sometimes they form stromatolites that look almost as though they might be fossil cabbages. Geologists thought for many years that these different forms were the remains of different species of plants. But more-detailed study showed that their characteristics all grade continuously into each other so it now appears that the various forms developed when the same kind of plant grew in different environments.

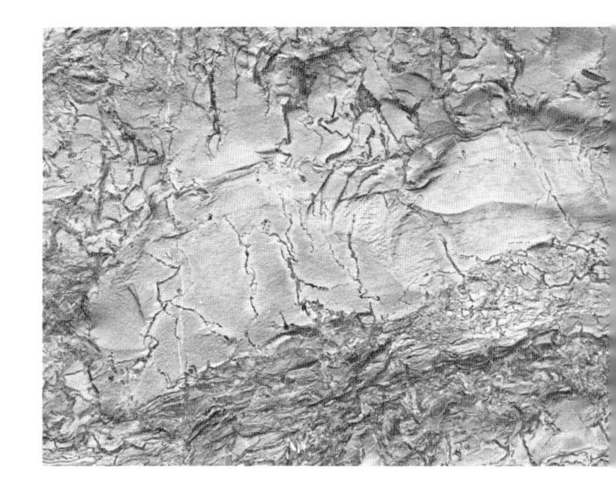

*Crusts of calcium carbonate that formed on blades of blue-green algae were crumpled by compaction of the mud after they had been buried. These "molar tooth structures" are believed to have formed where the algae grew in quiet water.*

*Outcrop of Altyn Limestone beside the parking lot of the Many Glacier Hotel. Thin beds of quartz sand grains stand out in relief on the weathered surface.*

Blue-green algae, like other green plants, use carbon dioxide and sunlight to create the starches and sugars in their tissues. When they grow under water, their use of carbon dioxide makes the nearby water alkaline causing precipitation of a crust of calcium carbonate on the plant. To avoid being smothered, the plant grows up through the crust and covers it with a new growth of algae. As this happens again and again, the plant causes precipitation of a nested series of thin calcium carbonate crusts. It is these, not the plant itself, that are preserved as fossils in the rocks.

Many geologists believe, for a variety of reasons, that the earth's atmosphere lacked oxygen and was very rich in carbon dioxide during much of the time when the Belt sedimentary rocks were deposited. Blue-green algae use carbon dioxide and release oxygen so their living in that ancient atmosphere must have helped convert it into the kind of air we breathe today. Probably we should regard these early forms of life with gratitude and respect instead of merely as curiosities.

*Altyn Formation*

The lowest, and oldest, sedimentary formation exposed in Waterton-Glacier Park is the Altyn Limestone (its lower part in Canada is called the Waterton Limestone) which outcrops along the base of the eastern front of the mountains. Altyn Limestone consists mostly of several hundred feet of white limestone that weathers to shades of tan on old surfaces. Its most distinctive feature is numerous thin layers of quartz sand which give the formation an internal structure of crisscrossing minor beds. Their pattern suggests that the entire formation was originally laid down as a beach deposit of sand-sized grains of the minerals calcite and quartz. The calcite grains have since welded themselves together to form solid limestone while the quartz grains have remained unchanged and still preserve traces of the original layering. A few layers of the Altyn Limestone contain stromatolites.

*Stromatolite head in the Siyeh Limestone near Logan Pass.*

### *Appekunny Formation*

Above the Altyn Limestone is the Appekunny Formation, easily recognizable on distant mountainsides as a broad band of greenish-gray rock from 2500 to 3500 feet thick. Most of the beds are green mudstones that weather to various shades of buff and dark gray-green on old surfaces but layers of red mudstone also exist and become abundant near the top of the formation. Beds of white sandstone, several of them thick enough to make conspicuous white stripes on mountainsides, occur throughout the formation. Surfaces of sedimentary layers in the Appekunny Formation frequently bear patterns of ripple marks and mudcracks showing that the original muds were deposited in an environment that was alternately wet and dry.

*Bedding surface in the Grinnell Formation showing a pattern of cracks formed as mud dried in the sun.*

### *Grinnell Formation*

As the frequency of red layers increases, the upper part of the Appekunny Formation passes transitionally into the overlying Grinnell Formation which is composed mostly of red mudstones. Layers of green mudstone and white sandstone also occur. Bedding surfaces within the Grinnell Formation are liberally marked by mudcracks, ripple marks, raindrop impressions, and other sedimentary features, indicating that the original muds were deposited in an environment that was alternately wet and dry.

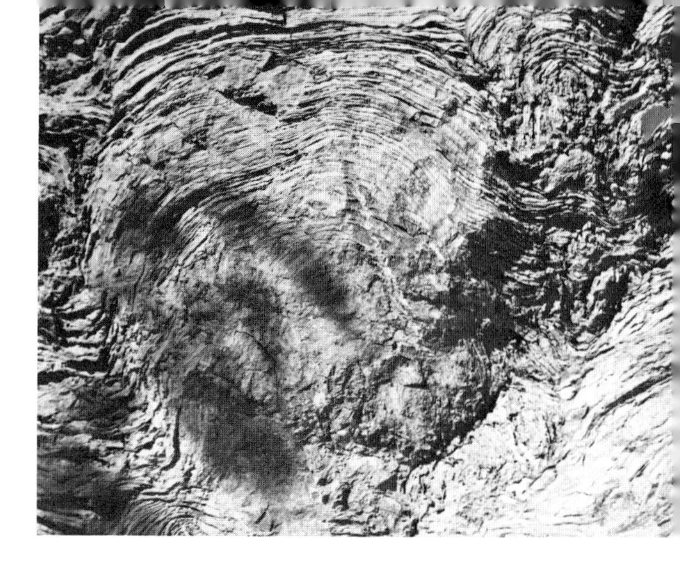

*Stromatolites (algal heads) exposed in a roadcut 0.2 mile below the tunnel west of Logan Pass.*

Red and green mudstones of the Grinnell and Appekunny Formations are nearly identical except for their color. Apparently the color variation is caused by a difference in the degree of oxidation of iron; green mudstones contain ferrous iron in the mineral chlorite whereas red mudstones contain ferric iron in the mineral hematite. Most available evidence suggests that the color variation is an original feature of the rocks and has something to do with the environment in which the muds were laid down over a billion years ago. It is possible to speculate on the meaning of the red and green colors but no one can be confident that he knows the answer. Perhaps the red muds were deposited in an environment that was becoming alkaline while it dried in the sun and the green muds formed in one kept fresh by rainfall.

*Upper surface of a layer in the Appekunny Formation showing a pattern of ripple marks formed by wave action in a shallow pool.*

### *Siyeh Formation*

Above the Grinnell Formation is the Siyeh Limestone, 2500 to 3500 feet thick and the most conspicuous formation in the Park. Organic matter trapped in the formation while it was deposited stains the fresh rock almost black but it weathers to various shades of tan and light gray. Siyeh limestone is a very strong rock, it generally forms bold cliffs which dominate much of the scenery in the higher parts of the Park.

Fossil blue-green algae are abundant throughout the Siyeh Limestone. Most often they are in the form of thin layers or crumpled leaf-shaped blades, but some beds contain well-developed stromatolites of all sizes (pictures on page 6). We can be sure that this limestone was originally deposited in very shallow water, otherwise all that plant life coult not have flourished in the sunlight. Very green and scummy mudflats must have existed here over a billion years ago when the Siyeh Limestone formed.

### *Shepard and Kintla Formations*

Above the Siyeh Limestone is the Shepard Formation and above that the Kintla Formation. These are composed of a variety of rock types but tan limestones and siltstones dominate the Shepard Formation and red mudstones the Kintla Formation. Both are rich in sedimentary features indicating deposition in an environment that was alternately wet and dry. Neither formation is much seen by visitors to the Park because most exposures are on remote mountaintops far from the road, wherever gray cliffs of Siyeh Limestone are capped by tan and red layers. However, both formations are prominent in the immediate vicinity of Logan Pass and along Highway 2 between Marias Pass and West Glacier.

*Salt casts on a bedding surface in the Kintla Formation. These form when cube-shaped crystals of salt growing in an evaporating pool make imprints in the muddy bottom. About natural size.*

*Pebbles of red and green mudstones make colorful stream gravels.*

# Igneous Rocks

Not long after the last bed of Siyeh Limestone had been deposited, while the mudstones of the lower part of the Shepard Formation were being laid down, molten magma rose from deep within the earth and squirted into the thick sequence of Precambrian sediments. Some of it cooled very slowly underground to form sills and dikes of coarse-grained, black diabase. The rest erupted onto the surface as lava flows that cooled quickly to form fine-grained, black basalt. Diabase and basalt are basically the same kind of rock; they differ from each other mainly in grain size, a matter of how much time was available for crystals to grow while the magma cooled.

Molten magma intruded between sedimentary layers cools to form a kind of intrusion called a sill, a layer of igneous rock sandwiched between layers of sedimentary rock. If the magma injects into fractures it cools to form dikes that cut steeply across the sedimentary layers.

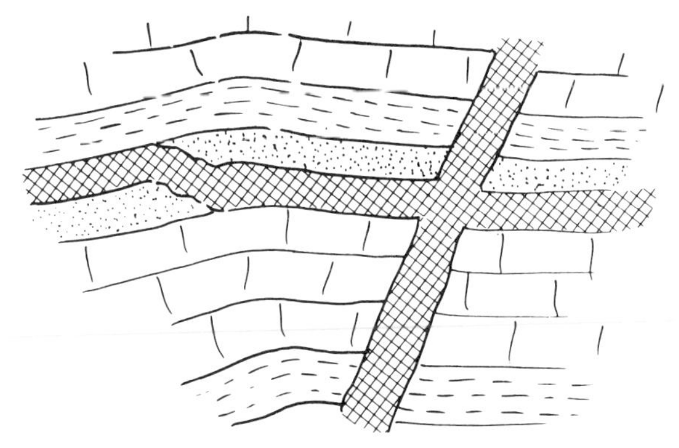

*A dike formed where magma squirted into a fracture cutting across sedimentary layers and a sill formed where it squirted out between the layers.*

## *Dikes and ore minerals*

Under the climatic conditions that prevail in Waterton-Glacier Park, diabase weathers to soil much more rapidly than does limestone. So the diabase dikes are rarely well-exposed but generally weather back to form deep grooves, most often noticed as lines of snow running steeply up the side of a mountain to a notch on the crest. The most-easily accessible good exposure of a large diabase dike is at the upper end of the tunnel along the Going-to-the-Sun Road west of Logan Pass (see photograph on page 68).

Wherever diabase intrusions have been emplaced in limestone it is commonplace to find that ore minerals were introduced along with the igneous rocks. Limestone adjacent to the diabase dikes in Waterton-Glacier Park contains small quantities of copper minerals which were the object of a brief mining rush at the turn of the century. Prospectors found mineralization in the Siyeh Limestone on the eastern side of the Continental Divide and then applied political pressure to get that area removed from the Blackfeet reservation and made available for mining. Efforts to develop the deposits failed miserably because there is not enough ore to be worth mining. Hikers in the Swiftcurrent Valley still encounter ruins of old mines now overgrown by trees. Nothing remains of the small town of Altyn that once stood, but never really prospered, beside Swiftcurrent Creek about a mile downstream from where now stands the Many Glacier Hotel.

View across Lake Josephine near Many Glacier. The Purcell Sill is the dark band of black diabase enclosed between bleached beds of Siyeh Limestone. Some of the notches on the skyline mark deeply weathered diabase dikes. National Park Service photo.

### Purcell Sill

Diabase sills often outcrop prominently in Waterton-Glacier Park because they are sheltered from the effects of weathering by their roof of limestone. A large sill approximately 100 feet thick, called the Purcell Sill, forms a black band in the upper part of the Siyeh Limestone. It is thinly bordered above and below by conspicuous white zones formed when heat from the magma drove the black organic material out of the adjacent Siyeh Limestone, bleaching it white and recrystallizing it to form marble. Look for the Purcell Sill where the sun lights the high, gray cliffs of Siyeh Limestone in the central part of the Park.

### Purcell flows

Some of the molten magma broke through all the layers of sedimentary rock and erupted onto the surface to form basalt lava flows, called the Purcell Flows, that now outcrop in the northern part of the Park. Good exposures are most easily accessible around Granite Park Chalet, north of Logan Pass, and in the western part of Waterton Park. Granite Park was named by someone who thought the black basalt was granite, a rock that does not occur in the Park.

Evidently the eruptions occurred during deposition of the lower layers of the Shepard Formation because the lava flows are buried within those strata. And the area must have been submerged, at least shallowly, because many exposures of the Purcell Flows show pillow structures. These form when a basalt lava flow erupts under water; still-molten magma within the flow repeatedly breaks through a rapidly-formed skin of solidly-chilled rock on the surface to form a series of bulging lobes about the size and shape of pillows. Dripping candle wax does the same thing on a much smaller scale.

### *Age of the igneous rocks*

Although no direct connection has been observed, it seems certain that the Purcell Sill was injected at the same time the Purcell Flows were erupted and that both were derived from the same magma. Analysis of radioactive elements contained in minerals of the Purcell Sill show that it crystallized about 1080 million years ago, probably the Purcell Flows erupted at the same time. If so, the lower part of the Shepard Formation was deposited about 1080 million years ago and all the rocks below that level must be older. All those above are, of course, somewhat younger. The radioactive date on the Purcell Sill is the best clue we have to the age of the Precambrian rocks in Waterton-Glacier Park.

*Red mudstones of the Kintla Formation cap peaks above Logan Pass. Danny On photo.*

# Paleozoic and Mesozoic Sedimentary Rocks

*Paleozoic rocks*

Geologists call the interval of time between the beginning of the Cambrian Period, about 600 million years ago, and the end of the Permian Period, about 225 million years ago, the Paleozoic Era. Shallow seas submerged the region that includes Waterton-Glacier Park several times during this long interval and thick deposits of sedimentary rocks, mostly limestones, were laid down. These formations make ranges of mountains north and south of the Park but are not exposed within its boundaries. Presumably they exist at depth in the Park, deeply-buried beneath other rocks and could be reached only by drilling.

*Mesozoic rocks*

During the latter part of the Cretaceous Period, between about 100 and 70 million years ago, the area of Waterton-Glacier Park was again shallowly-submerged. This time an inland sea stretched far to the east from a western shoreline in the vicinity of the present Rocky Mountains. Thick deposits of sedimentary rock, mostly black mudstones and muddy sandstones, accumulated along the shores and beneath the waters of this inland sea. These outcrop today along the eastern side of Waterton-Glacier Park. They contain numerous fossils of small marine animals that lived during Cretaceous time, modest contemporaries of the great dinosaurs whose bones are occasionally found in the same rocks.

Most beds of Cretaceous sediments in Waterton-Glacier Park are not thoroughly hardened into rock, mostly because they have never been deeply buried. Bold outcrops are uncommon. These rocks support a distinctive landscape of gently-rolling small hills covered by a patchy forest of gnarled aspens and stunted conifers.

*Layers of brown sandstone are typical of the Cretaceous marine sediments that outcrop along the eastern front of Waterton-Glacier Park. These are beside the highway south of Kiowa.*

Cretaceous sedimentary rocks exposed along the eastern part of Waterton-Glacier Park are tempting to the petroleum geologist. Black marine mudstones, rich in organic matter, and interbedded with layers of sandstone are precisely the kind of sedimentary accumulation that frequently generates deposits of petroleum. Several good oilfields produce from these rocks in Alberta and Montana and prospecting activities have extended to the eastern boundary of the Park.

*Subdued topography in foreground is eroded on soft sedimentary rocks deposited in a shallow Cretaceous sea. Steep mountains in the background are underlain by harder Precambrian sedimentary rocks above the Lewis Overthrust. Melting ice left the large boulders.*

Montana's first oil wells were drilled in the Cretaceous rocks beneath the Swiftcurrent Valley during the early years of this century before the Park was established. They produced very small quantities of oil from a depth of about 500 feet. Similar activity took place at about the same time along Cameron Creek in Waterton Park. Natural oil seeps have been reported from a number of places in the Park including several in the North Fork Valley.

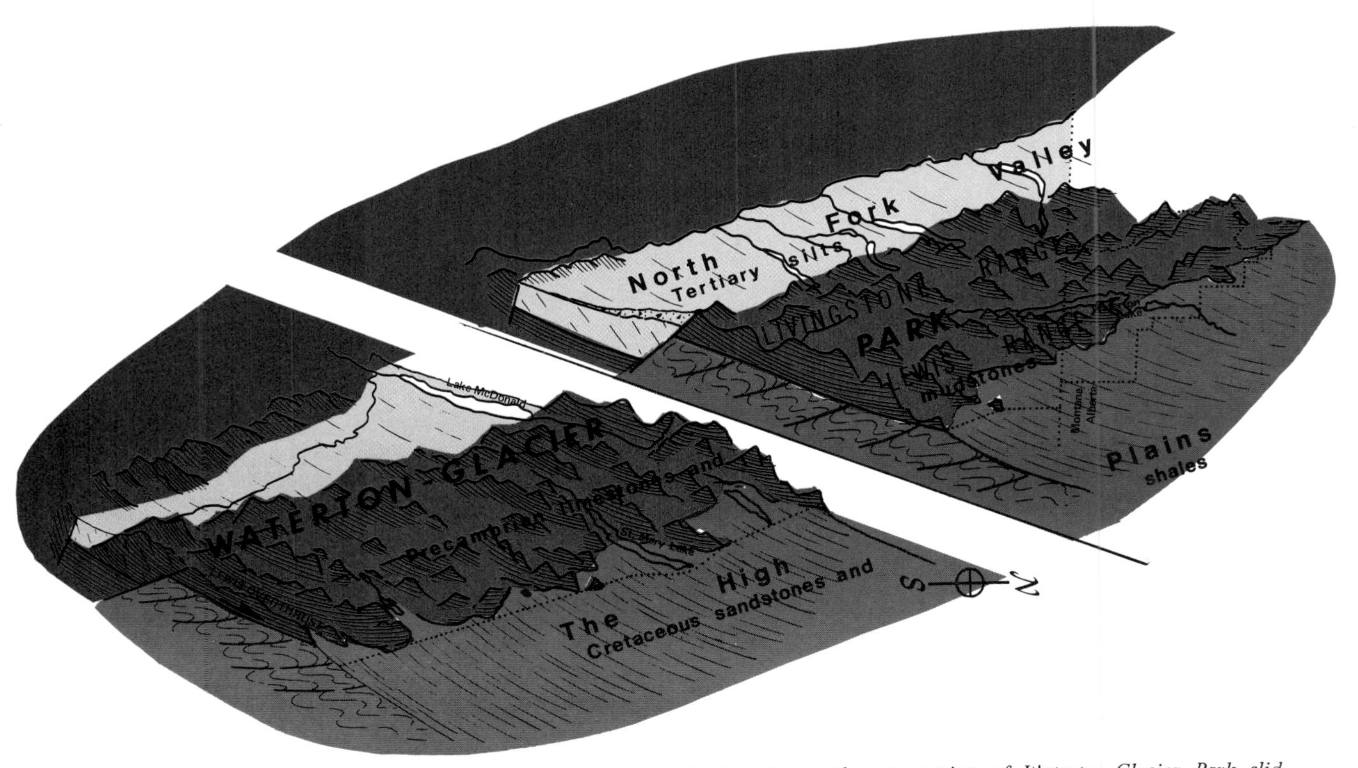


*The Precambrian slab that forms the mountains of Waterton-Glacier Park slid eastward leaving the North Fork and Flathead valleys as pull-apart openings.*

# The Lewis Overthrust Fault - The Beginning of Mountains

Geologists normally expect to find sedimentary rocks stacked in the order they were deposited with the oldest layers at the bottom of the pile and the youngest on top. In Waterton-Glacier Park this expectable order of rock layers is dramatically reversed; Precambrian sedimentary rocks deposited more than one billion years ago are now on top of Cretaceous sedimentary rocks laid down only 70 million years ago! Apparently a slab of Precambrian sedimentary rock several thousand feet thick slid eastward some tens of miles over the much younger Cretaceous sedimentary rocks. The surface on which the sliding movement occurred is called the Lewis Overthrust Fault. Geologists have marvelled at this situation, and attempted to understand it, ever since they first recognized it in the years near the beginning of the twentieth century.

For many years geologists argued that a strong force from the west must have pushed the slab of Precambrian sedimentary rock eastward onto the plains. But this vision poses problems: pushing such a long, thin slab of rock from behind without crumpling it is an undertaking similar to pushing a large carpet across a floor without wrinkling it. The rocks, like the carpet, are not strong enough to withstand the necessary force without folding, and the friction in the sliding surface makes it difficult for movement to occur. The Precambrian slab in Waterton-Glacier Park is not crumpled, neither is there evidence in the area west of the Park of any geologic event that could have exerted a strong eastward push. So it is difficult to imagine how this enormous slab of rock was pushed into its present position.

Now most geologists are convinced that overthrust faults do not move in response to a push from behind but instead slide downhill under the pull of gravity. This eliminates the problem of understanding how the overthrust slab could withstand a force without crumpling because every particle of rock feels the tug of gravity individually.

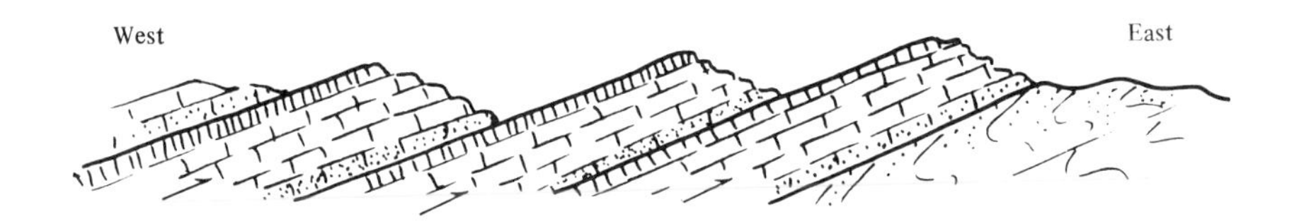


*Large slabs of rock slid into a stack to form the geologic structures of most of the front of the northern Rockies, including that of part of Waterton Park.*

The problem of overcoming friction in the sliding surface of the fault still remains. If water in the pores of the rock is under sufficient pressure, it will support the rocks above enabling them to move with very little friction. This might have happened in the Cretaceous rocks but would have been most unlikely in the much older Precambrian formations because they were probably much too dry by the time movement occurred on the Lewis Overthrust fault. Some geologists contend that the Precambrian rocks moved because certain layers were weak enough to flow like a viscous fluid carrying the rocks above. They imagine the rocks behaving somewhat like a layer cake filled with soft frosting.

Thousands of feet of sedimentary rocks had accumulated in the region of Waterton-Glacier Park by the end of Cretaceous time. Layer after layer of sandstone, mudstone and limestone deposited over a period of many hundreds of millions of years while the region was a level plain flooded at times by shallow seas. The quiet years ended about 70-90 million years ago, during Cretaceous time while giant dinosaurs still roamed the earth.

Processes operating deep within the earth slowly raised the continental crust of western North America, with its burden of sedimentary rocks, several thousand feet vertically upward forming a broad crustal arch that had its crest about 100 miles west of Waterton-Glacier Park. Thousands of feet of interlayered limestone, sandstone, and mudstone slowly slanted eastward as uplift proceeded. Tilting such a thick pile of sedimentary rocks, some layers strong and others weak, is somewhat like tilting a thick stack of heavily-buttered pancakes. In both cases the entire stack will eventually begin to slide downward on one or more of the weak layers. That is probably what happened about 50-60 million years ago all along the eastern front of the northern Rockies.

Somewhere deep within the pile of sedimentary rocks a weak layer yielded under the strain and the rocks above began to glide slowly eastward down the flank of the broad regional arch still rising in the earth's crust. The huge slab of Precambrian sedimentary rock that is now Waterton-Glacier Park slid an unknown distance eastward, probably more than 30 miles. It moved at least ten miles across the soft sands and muds deposited in the shallow Cretaceous sea that had receded only a few million years previously.

As the mass of Precambrian sedimentary rock slowly slid eastward, large gaps opened behind it. The same sort of thing happens when slabs of snow slide down a roof leaving gaps where the moving snow detached itself from the stationary mass that remained behind – or the upper end of a glacier pulls away from the mountain at its head. The North Fork Valley, along the western margin of Waterton-Glacier Park, appears to be such a gap pulled open behind the Precambrian slab as it moved eastward on the Lewis Overthrust. It is approximately ten miles wide, probably not enough to account for all the displacement on the Lewis Overthrust fault. In order to find another gap to account for more displacement we must go farther west to the broad Flathead Valley and suggest that the Whitefish Range, west of the North Fork Valley, may also have slid eastward.

Early geologists interpreted the North Fork Valley quite differently, arguing that it is a downdropped block of the earth's crust. Rocks in the southern end of the valley, where the bedrock valley floor is exposed, do suggest that the valley floor has dropped vertically downward. But there are reasons to believe that the floor of a pull-apart valley might drop so the two explanations are not incompatible.

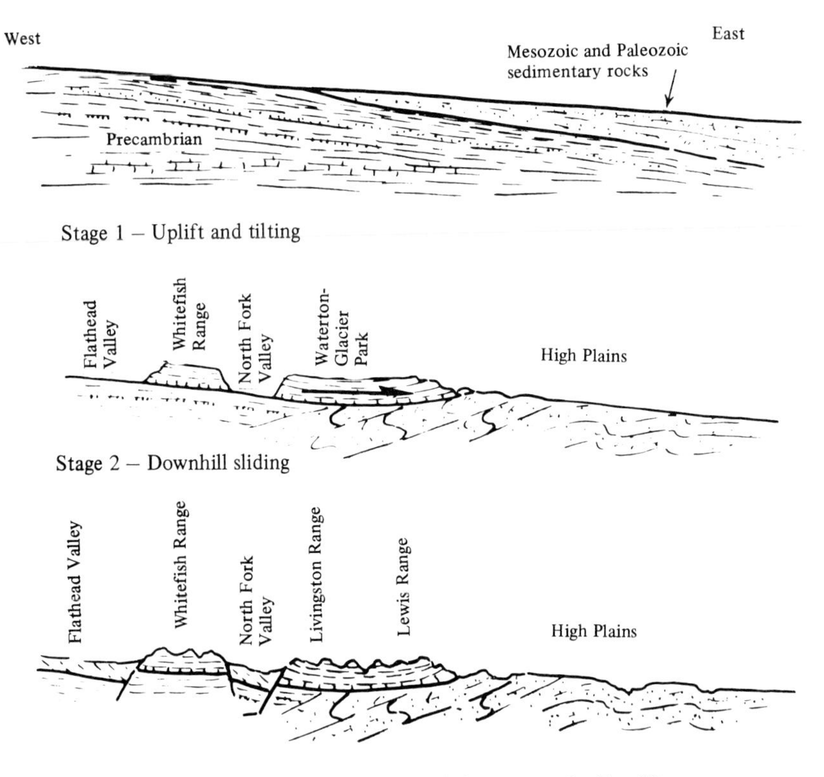


*Sequence of events as a slab of Precambrian sedimentary rocks slid eastward over younger rocks to form the mountains of Waterton-Glacier Park. The slab is folded to form a broad trough with edges that make the Lewis and Livingstone Ranges.*

Mountain ranges north and south of Waterton-Glacier Park also consist of large slabs of rock that slid eastward as the earth's crust arched up to the west. But these ranges do not consist of a single large slab. Instead, they are made of many smaller slabs stacked on each other like a row of fallen dominoes. This kind of structure is also found in part of the eastern portion of Waterton Park (see diagram on page 22).

# Preglacial Erosion

Movement on the Lewis Overthrust Fault ceased about 50 million years ago, very early in the Tertiary Period, and erosion has been carving the overthrust slab into mountains ever since. Most of the landscape in Waterton-Glacier Park was sculptured by the great ice age glaciers of the past three million years. But enough remains of the preglacial landforms to make them significant elements in the modern landscape.

### *High Plains Surface*

Visitors traveling east of the Park see extensive smooth surfaces, largely covered by wheat fields, that cap the highest elevations on the plains and slope almost imperceptibly downwards to the east. Roadcuts immediately under this upland surface expose beds of coarse, water-rounded gravel.

It is easy, while driving across the plains, to mentally reconstruct the former extent of this old surface by imagining the surviving remnants connected from one side of a modern stream valley to the other. Evidently the high plains surface was continuous over the entire region before it was dissected by streams.

### *Dry Climate – Rainsplash Erosion and Valley Filling*

Extensive smooth surfaces underlain by water-transported gravels form today wherever the climate is so dry that only scattered desert plants can grow. So it seems reasonable to conclude that the similar high plains surface also formed under such conditions. It is a relic of a time when the climate of the northern Rockies was much drier than it is today.

Where plants are few, raindrops strike directly onto bare soil, splashing it into the flow of surface water that sweeps it away. That is why deserts have a high rate of soil erosion even though they don't receive much rain. Most of the rain that falls in such regions runs off the surface, instead of soaking into the ground, so there are frequent flash floods. These sweep across the countryside creating extensive smooth plains veneered by deposits of water-transported sediment. Another characteristic of desert regions is the tendency for thick deposits of water-transported sediment to accumulate in the valleys where the transporting streams dry up for lack of water.

While flash floods were grading the high plains surface east of Waterton-Glacier Park, the North Fork Valley on its western side was filling with sediment. No one has ever figured out how thick these deposits are but other, similar, valleys in the region are known to contain a thickness of between 2,000 and 3,000 feet of such sediment. Some of the oldest layers exposed in the North Fork Valley contain fossils of animals and plants that are known to have lived during early Tertiary time, about 50 million years ago. Clearly, the valley had already pulled open behind the Lewis Overthrust Fault by that time.

Marshes are common in undrained desert valleys and must have existed in the North Fork Valley while it was receiving sediment. Accumulations of plant remains were buried in the valley-fill sediments and became beds of low-grade coal. Some of these were mined, on a small scale, near the beginning of the twentieth century. But the deposits are small and their quality poor so it seems unlikely that they will ever support large mines.

By the time the first ice age began, about 3 million years ago, the mountains of Waterton-Glacier Park had been reduced to rather modest relief. To the east, they sloped away on the high plains surface which is more than 1000 feet above the elevation of present stream valleys. To the west, they rose above the North Fork Valley which was filled then with a much greater depth of valley fill than it now contains. And the mountains must have been angular like all desert mountains, their slopes furrowed by numerous gullies.

*Remnants of the high plains surface, cut by a modern stream valley, are silhouetted in this view looking west.*

# Ice Ages and Glaciers

At least four major ice ages, separated in time by much longer interglacial periods, have spread enormous glaciers over wide areas of the northern continents during the last 3 million years, the Pleistocene Epoch. The last ice age ended, and the present interglacial period began, approximately 10,000 years ago.

Most of the present landscape of Waterton-Glacier Park remains today essentially as it was left by the ice. Reshaping by modern processes of erosion during the last 10,000 years has added some small details to the major work done by the ice.

*Causes of Ice Ages*

Extensive glaciers have not existed during most of geologic time. Geologists have puzzled for more than a century over the riddle of why they formed, and then melted, at least four times during the last 3 million years. Although numerous theories have been proposed, the debate continues, still without a final answer. But there are some partial answers and some interesting ideas.

In order for glaciers to exist, more snow must fall in winter than the summer sun can melt. So a climatic change in the direction of either wetter winters or cooler summers could cause glaciers to form. Abundant evidence indicates that ice ages were times of very wet climates, but very little evidence suggests that their climates were greatly colder than the one we now have. So the explanation for the cause of ice ages must involve a change to wetter conditions. If a new ice age were to start now, the change in weather would benefit manufacturers of umbrellas and galoshes more than those who make fur coats. Many areas now arid would receive ample water while certain other regions now productive would be buried under a growing burden of snow and ice.

In order for the climate to become wetter, more water must be evaporated from the oceans into the atmosphere. One popular theory suggests that this might happen if the thin crust of floating sea ice in the Arctic Ocean were to melt. Increased evaporation in the Arctic would then convert the polar air masses, now dry, into moisture-bearing winds bringing such heavy winter snows that ice would accumulate from one year to the next. A climatic change or deliberate modification that improves navigation in the Arctic might also unleash a new ice age on the industrial nations of the northern hemisphere.

Not all geologists agree that this is the cause of ice ages so a lively debate still continues. However, most geologists do agree that the cause of ice ages, whatever it may have been, is probably still operative and that more ice ages are in prospect for the future. Few experts would venture to predict whether the next ice age can be expected in the near or distant future. But many who study the subject are seriously concerned that human intervention with the environment, especially the arctic environment, may prematurely precipitate the next ice age.

### *Wet Climates and Stream Erosion*

The onset of the first ice age, about 3 million years ago, ended many millions of years of desert conditions. Ever since then rainfall has been sufficient to enable streams to maintain a steady flow to the sea and plants to spread a dense cover of green over mountains that had been nearly bare. Now the soil soaks up rainfall and stores it instead of shedding it as runoff.

Where muddy flashfloods had swept across the countryside after every heavy rain, perennial streams now carry a steady flow of clear water and erode their channels deeper dissecting the old high plains surface. Now, after 3 million years, they have cut their valleys more than 1000 feet below that surface leaving its remnants stranded high on drainage divides. And the North Fork of the Flathead River began the long job of carrying away valley-fill sediment that had accumulated in the North Fork Valley during the time when there was no drainage. Much sediment still remains but enough has been removed to lower the valley floor about 1000 feet making the mountains seem that much higher. The jagged peaks and dramatically yawning valleys of the modern landscape were sculptured by the great ice age glaciers pouring slowly down channels originally started millions of years before by desert flash floods.

Here and there in Waterton-Glacier Park high benches still remain above the present valley floors to record the level they had before interglacial streams and ice age glaciers began their long job of valley deepening. One of these, Granite Park, is visible about 1500 feet above the present floor of McDonald Creek a few miles north of Logan Pass. We owe at least that much of the vertical scenery in the Park to erosion since the onset of the ice ages. Such scattered high benches, the high plains surface, and the sedimentary deposits in the North Fork Valley are all that remain of a landscape that took about 40 million years to form and existed as recently as 3 million years ago.

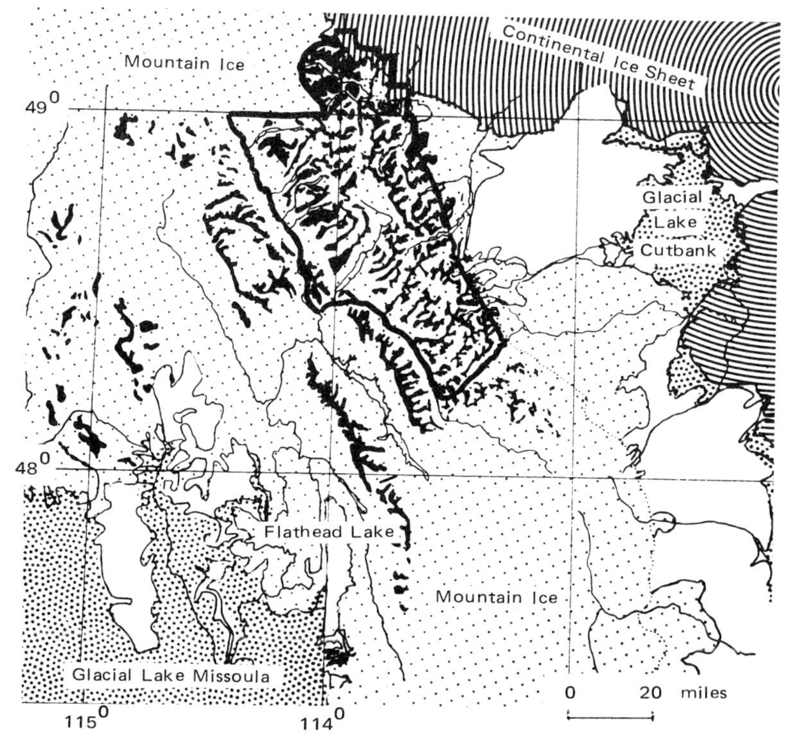


*Approximate maximum extent of glacial ice in and around Waterton-Glacier Park during the last ice age. Black areas are above ice. Partly after Alden, 1932, and Richmond et. al., 1965.*

### *Extent of Glaciation*

During the last ice age, up to about 10,000 years ago, the central part of Waterton-Glacier Park was virtually buried under a massive ice cap that poured rivers of ice down several major valleys. Smaller glaciers filled the other mountain valleys of the Park. Ice filled the lower parts of the North Fork Valley as well as the much larger Flathead Valley farther west. Mountain glaciers poured eastward onto the high plains where they met the edge of the great continental ice sheet.

Most evidence of earlier glacial periods was obliterated by the glaciers of the last one. But enough remains to make it clear that the most recent ice age was far from being the greatest. Earlier ice ages completely filled the North Fork Valley with ice and covered all but the highest peaks in the mountains of the Park. Glaciers flowing east onto the plains lapped much higher on the hillsides, some 500 to 1000 feet higher than the level of glaciers during the latest ice age.

During all the glacial periods ice also flowed radially outward from a large region of accumulation in northern Ontario and Manitoba to cover most of Canada and much of the United States north of the Ohio and Missouri Rivers. This continental ice reached southwestward into Alberta extending at times as far as the eastern side of Waterton-Glacier Park. Ice flowing eastward onto the plains from the mountain valleys of the Park met that flowing southwestward from central Canada.

*Smoothly-scoured and grooved surface cut on suncracked mudstones of the Shepard Formation by a small glacier at the base of Mount Clements. Small pebbles left scattered by the melting ice were the tools it used to carve the bedrock.*

*Hidden Lake occupies a cirque near Logan Pass. Ragged bedrock surfaces were quarried by moving ice. Marion Lacy photo.*

## *Glacial Erosion and Deposition*

Soft snow begins to change to granular ice very quickly, usually within a few days. The first stages of this change are familiar to anyone who postpones shoveling snow off his front walk for several days after a storm. Visitors to the Park who stop to examine midsummer 'snowbanks' quickly find that they are made of granular ice, not snow. Meltwater trickles into these during warm weather wetting the surfaces of ice grains so they will weld themselves into a solid mass when the weather turns cold.

As ice continues to accumulate, it eventually reaches a thickness sufficient to cause the solid ice at the bottom to flow. A glacier is born. Glacial ice moves almost as though it were very heavy syrup or tar. The exact manner of ice flowage is complex involving a combination of such processes as recrystallization and deformation of ice grains under pressure, shearing movements within the ice, and perhaps sliding of the entire glacier slowly over its rocky bed. Glacial flow begins when the ice reaches a critical thickness somewhat greater than 100 feet. The rate of flow varies between a few feet and several hundred feet per year depending upon the temperature, the thickness of the ice and the slope of the surface. It is often difficult to determine exactly when an accumulation of ice passes beyond being merely a large snowbank and becomes a small glacier. Many of the glaciers in the Park are near that difficult borderline so estimates of their number vary considerably. Thick snow may also locally creep slowly downslope contributing to the difficulty of distinguishing precisely between large snowbanks and small glaciers.

Glacial ice keeps flowing away from its highest area of accumulation until it reaches a point where the climate is warm enough to melt the ice as fast as it advances. The margin of a stable glacier always marks the point of balance between advance and melting of the ice. Climatic changes which increase the supply of ice or the rate of melting will cause the glacier to advance or recede, respectively.

Glaciers erode their beds in several ways. Ice freezes tightly onto soil and loosened pieces of bedrock and then drags them away as it flows downhill. Surfaces quarried in this way are left as ragged exposures of bedrock after the ice has melted. The floors of many high valleys in the Park are good examples. Rocks and particles of soil embedded in glacial ice are dragged across the rock surface rasping it away. Surfaces eroded this way are left smoothly-polished except for deep scratches gouged by larger chunks of rock. Processes of weathering rather quickly destroy glacially-polished surfaces so these are best observed near retreating glaciers or where protective soil has recently been removed from the rock.

*Air view looking west up Two Medicine Valley. Glaciers carved cirques into mountain peaks leaving jagged spires and knife-sharp ridges. National Park Service photo.*

*Glacially-scoured valley walls loom above the upper Many Glacier area.*

As glaciers move downhill they pull away from the rock wall at the head of the valley, opening a large gap between ice and bedrock. Meltwater pours into this opening and freezes, firmly fastening the ice deep in the glacier to the rock. Continued downhill movement of the glacier pulls the ice away from the headwall again, plucking away chunks of the bedrock. Thus glaciers undercut the bedrock headwalls of their valleys and erode their bases deeper into the mountain. After the ice has melted, the quarried heads of glaciated valleys are left as great cliff-walled basins, called cirques, hollowed out of the mountains as though by a giant at play with an ice cream scoop.

Most prominent mountain peaks in Waterton-Glacier Park were eroded by several glaciers that scooped cirques out of their summits. Now that the ice has melted, all that remains of what was once a broadly-rounded mountain is a pointed spire with sides that drop away into deep cirque basins.

Glaciers rarely create valleys of their own but gouge their way down older stream valleys, planing away the curves. After the ice has melted, glaciated valleys are left as straight troughs with very steep sides and broad floors filled with alluvial gravels.

*Birdwoman Falls as seen from Going-to-the-Sun Road west of Logan Pass. The main valley was greatly deepened by a very large glacier so the tributary now enters over a waterfall from a hanging valley.*

Many of the long, unobstructed views up the valleys of Waterton-Glacier Park would be quite impossible had they not been glacially straightened.

Where two glaciers meet, their upper surfaces are at the same level. But a large glacier is much more effective at deepening and gouging the rock walls of its valley than is a smaller one so their lower surfaces will not be at the same level. This becomes obvious after the ice has melted revealing the valley walls and floor that it eroded. Tributary streams must enter the greatly deepened main valley by dropping into it over a waterfall. Such hanging tributary valleys are one of the most characteristic features of a glaciated mountain landscape.

Where glaciers scour adjacent valleys, the ridge between them is gouged away on both sides and may be reduced to a narrow blade of rock with a sharp and ragged crest. These knife-like ridges are typical of glaciated mountain ranges; sometimes they are called "cleavers" but the French word "arete" is more frequently used. The Garden Wall north of Logan Pass is an excellent example familiar to most park visitors. Its top is so thin that it actually has a hole through it; people who enjoy midsummer twilights from the Many Glacier area like to wait for the moment when the setting sun blazes briefly through the hole in the Garden Wall.

*View down McDonald Creek, a valley deeply gouged and straightened by a large glacier. Tributaries enter from hanging valleys. Marion Lacy photo.*

Gouged valleys, cirques scooped into mountain tops, knife-edged ridges and peaks carved away to gnarled pinnacles, all combine to give glaciated mountain ranges their jagged appearance. Had it not been for the glaciers, these mountains would be smoothly rounded like the occasional south-facing slope that escaped glacial erosion. Many people mistakenly suppose that young mountains are angular and old ones rounded. Whether mountains are smooth or angular depends upon the kind of rock and the processes of erosion that carved them and not upon the age of either the mountains or of the rocks they contain.

Material eroded in one place must be deposited in another. Rock eroded by glaciers from the upper parts of their valleys is deposited lower down to make easily recognizable landforms. Most glacial deposits can be broadly separated into two general categories: those composed of till, material laid down directly by glacial ice; and those composed of outwash, material laid down by running water coming from melting glacial ice.

*View southwest toward Sperry Glacier in the distance, center. The sharp ridge in the center foreground is an arete separating the cirque holding Hidden Lake, right; from that heading the St. Mary Valley, left. National Park Service photo.*

*Rugged, glacially-carved landscape in the south-central part of the Park. U.S. Air Force photo.*

Glaciers are capable of carrying debris of all sizes ranging up to boulders as big as a house. So deposits of glacial till usually consist of a disorderly mixture of mud, sand, gravel and boulders all dumped indiscriminately in a heap, like dirt scraped and dumped by a bulldozer.

Moraines are landforms made of glacial till. Frequently these are smoothly-plastered to the walls of the lower part of a glaciated valley to form "lateral moraines," so-called because they follow the sides of the glacier. Till usually makes fertile soil and almost always holds water so lateral moraines generally support a lush growth of forest. Visitors to the Park can often recognize them by observing the pattern of forest distribution on valley walls. A nearly-horizontal line commonly separates heavily forested lower slopes once buried under ice and now covered by lateral moraine from much more barren upper slopes where trees struggle to grow on bare rock.

*View across McDonald Creek from the "loop" in the Going-to-the-Sun Road. Lateral moraine plastered to the valley wall by glacial ice has slumped leaving a large landslide scar now grown over by lighter-colored vegetation.*

*Glacial till exposed in a roadcut through a moraine between St. Mary and Babb. Material of all sizes is jumbled together with no sorting or layering. Notice the scratched surface on the large boulder.*

Glaciers also dump till along their snouts to form deposits which remain as ridges after the ice melts. These are called terminal moraines if they mark the farthest advance of the ice and recessional moraines if they record a position occupied for a period after the ice had begun to retreat. Both terminal and recessional moraines appear as hummocky, tree-covered ridges swinging across a valley in a curving loop to connect the lateral moraine on one side to that on the other.

Terminal or recessional moraines sometimes function as natural earth-fill dams, impounding the drainage of a valley to form lakes or swamps. Some of the large lakes at low elevations around the margins of Waterton-Glacier Park are behind moraine dams.

*The ridge is a moraine in Waterton Park deposited between the Waterton and Blakiston glaciers where they merged. Blakiston Brook has sliced through it exposing a nice cross-section of glacial till.*

*Glacial outwash spread as a fan-shaped deposit at the mouth of Blakiston Brook. Waterton Lake is dammed by these gravels and those of Sofa Creek across the valley. Canada Parks photo.*

Much of the debris eroded by glaciers is deposited as outwash by meltwater streams. Running water, unlike ice, is sharply limited in the size of material it can transport; so outwash deposits are well-sorted into separate layers of sand, gravel, and silty mud. In this respect, glacial outwash resembles any other stream deposit.

Meltwater never plasters outwash onto hillslopes or piles it into morainal ridges the way ice deposits till. Instead, glacial outwash floors stream valleys and spreads over flat areas as gently-sloping deposits with smooth upper surfaces. Occasionally these are pitted by small kettle ponds, marking places where large blocks of ice buried in the outwash melted to form collapse depressions. All the large valleys in Waterton-Glacier Park are floored in their lower reaches by outwash deposits. Where these were laid down unevenly lakes may be trapped in the low places. The St. Mary Lakes, for example, are impounded in outwash deposits swept into the St. Mary Valley from large tributaries. Sometimes outwash deposits are laid down directly on the ice and then left as hummocks or small ridges after the ice melts. Deposits of this kind are especially conspicuous along the eastern side of Waterton Park.

*Kettle ponds dot the hummocky, morainal landscape left by the continental ice sheet at the edge of the plains. Canada Parks photo.*

# Modern Processes of Erosion

Melting glaciers leave behind them a bleakly dramatic landscape of craggy mountains deeply-gouged by steep-walled valleys. Jagged mountain tops that had been above the ice are left littered with angular debris formed as the bedrock was shattered by water freezing within cracks. Moving glaciers scrape away soil and loose rock so sidewalls of valleys that had been beneath the ice are left as scrubbed rock. In places bedrock is exposed in the valley floors as well but these are mostly buried in deposits of muddy debris left where it dropped as the ice melted.

Now that glaciers have almost entirely disappeared from Waterton-Glacier Park, other processes of erosion have taken over and begun the work of reshaping the landscape left by the ice. These are the processes that visitors to the Park can watch in action.

*Glaciers*

Despite the name, glacial erosion is no longer an important process in Waterton-Glacier Park. The mighty rivers of ice that carved the peaks and gouged the canyons just a few thousand years ago are gone. All that exist today are a few dozen small glaciers, most of them hardly more than large snowpatches, that cling to a precarious survival deep in the shadows of some of the higher peaks. These are not remnants of the Pleistocene glaciers, they formed much more recently. But they are genuine glaciers and they do illustrate, in miniature, the processes of glaciation.

*Sperry Glacier, a shrunken remnant of its former self, nestles in the shade of Gunsight Mountain. View looking southwest from a point above Logan Pass. National Park Service photo.*

During the early years of Glacier Park, W.C. Alden of the United States Geological Survey counted 90 glaciers within its boundaries. Most of these are shown on the large topographic map of the Park drawn during those years and still in use, a masterpiece of topographic surveying with plane-table and alidade. But the twentieth century has dealt harshly with glaciers, all of those in the Park have shrivelled and many have disappeared entirely. It seems unlikely that many more than half of those Alden counted survive as active glaciers today. Such numbers are somewhat uncertain because most of the glaciers in the Park are near the difficult borderline between small glaciers and large snowpatches.

None of the glaciers in the Park are accessible without hiking although several are visible from the road. Grinnell Glacier, at the head of Swiftcurrent Valley, is the largest in the Park and is also the easiest to reach. Sperry Glacier, at the head of Avalanche Creek, was until recently the largest in the Park and is visited by a great many people. Most of the others are smaller and rarely visited.

Sperry and Grinnell Glaciers have for many years been the subject of systematic study jointly conducted by the Park Service and the Geological Survey. Field parties visit both glaciers every year during late summer to make careful and detailed observations of the glacier's size, rate of movement and growth or shrinkage. Precise surveys are run along profile lines across the ice and of the ice margin. These indicate the size of the glacier and whether it has been growing or shrivelling since the year before. Marked poles are set to known depth within auger holes drilled into the ice to indicate the extent to which ice is being added by snowfall or lost by evaporation and melting. Large boulders, too big to be moved by human beings, are marked with painted numbers and their position surveyed from year to year to determine the rate of ice movement. Photographs are taken every year from specified vantage points to provide permanent records of the details of each glacier's appearance. Several of these photographs are used as illustrations in this book. These data, as they accumulate over the years, provide an excellent portrait of Sperry and Grinnell Glaciers which probably reflects the behavior of most of the other glaciers in the Park.

Sperry Glacier was first explored in 1897 by a party led by Dr. Lyman Sperry. A few years later, in 1901, it was estimated to cover an area of 810 acres making it the largest glacier in the region of the Park. By 1938 it had shrunk to 390 acres and in 1950 to 305 acres making it slightly smaller than Grinnell Glacier. By 1960 Sperry Glacier was reduced to an area of 287 acres. Since then it has continued to shrink but at a much slower rate so that it now appears to be stabilizing at a smaller size. During the same period of time the glacier also lost overall thickness, thinning to between 400 and 500 feet. Sperry Glacier moves very slowly, even for a glacier, averaging between 12 and 20 feet per year in the center section where rate of movement is probably greatest.

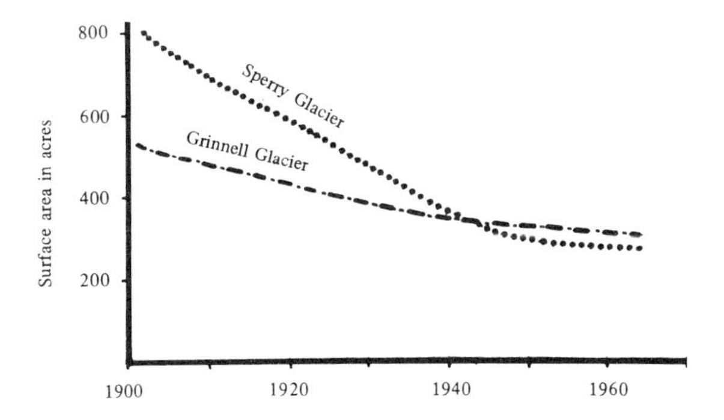


*Decline of Sperry and Grinnell Glaciers during the twentieth century. Graph compiled from data in open file reports by the U.S. Geological Survey and the National Park Service.*

*Glacially-scoured ridge at Many Glacier.*

Grinnell Glacier was first explored in 1887 by a party led by George B. Grinnell. In 1901 its area was estimated to be about 525 acres. By 1937 this had diminished to 385 acres, to 314 acres by 1950 and to 300 acres by 1960. Like Sperry Glacier, Grinnell Glacier shrank rapidly during the first half of the twentieth century and now appears to be stabilizing at a smaller size as it continues to shrink more slowly. Maximum thickness for Grinnell Glacier is probably between 400 and 500 feet, about the same as for Sperry Glacier. But Grinnell Glacier moves a little more than twice as fast as Sperry Glacier, between 30 and 50 feet per year near the center where the rate is probably greatest.

*Front of Sperry Glacier in late summer. Layers of dirt separate annual accumulations of snow in the source area. National Park Service photo.*

*Firn line separates white snow of the previous winter, background, from old, dirty ice, foreground, on Sperry Glacier. Gunsight Mountain on the skyline, right. National Park Service photo.*

Glaciers accumulate snow in their upper reaches and then flow downhill until they reach a level where the rate of melting balances the rate of advance. Very small glaciers, such as those in the Park, exist in precarious balance because the area in which ice accumulates is separated from the head of the glacier by a distance of as little as a few hundred yards. Very small fluctuations in climate can cause large percentage variations in the size of such glaciers making them sensitive indicators of climatic change. Obviously, the dramatic shrinkage of glaciers during this century must reflect either an increase in summer melting or a decrease in winter snowfall, or perhaps some combination of the two.

Several precipitation gauges are operated near Sperry and Grinnell Glaciers. Their readings differ over a considerable range because there is great local variation in snowfall in such mountainous terrain but the totals are mostly in the range from 100 to 150 inches of precipitation per year. Continued accumulation of climatic data will eventually make it possible to establish exactly what kind of climatic change caused the drastic shrinkage during the first half of the twentieth century.

Late summer, when the annual snowmelt is about completed, is the most interesting time to visit a glacier. Then there is a sharp line, called the firn line, that separates the lower part of the glacier where dirty ice is exposed from the upper part where white snow that fell the previous winter still survives. The part of the glacier above the firn line is the area where annual snowfall exceeds melting and new ice is accumulating. An object dropped on the ice in this area will be buried deeper each succeeding year as new ice accumulates above it. Meanwhile, it will move slowly downslope as the ice flows and will eventually pass beneath the firn line. Once it is downslope from the firn line it will again approach the surface – still moving downhill with the ice – as layers of ice are melted from the surface of the glacier each summer. Rocks melting free of the ice near the toe of the glacier during late summer may have fallen on the head of the glacier hundreds of years ago.

Glacial movement takes place at ice depths greater than 100 feet where the pressure exerted by the weight of the ice above is sufficient to make that below flow. Ice above the level of flowage deforms by fracturing as it is carried passively along on the flowing ice beneath, the brittle surface layer of ice stretches over humps or around bends by opening cracks called crevasses. There is almost always a large crevasse, sometimes called the "bergschrund," at the uppermost part of the glacier separating ice from bedrock. Crevasses can be dangerous, especially during early summer when they are likely to be concealed under a blanket of new snow. They extend through the surface layer of brittle ice down to the zone of flowage where they close.

*Crevasses in Grinnell Glacier at the foot of Mount Gould. Rock debris visible in the ice is destined to become part of a moraine. National Park Service photo.*

*This cobble of mudstone still bears the scratches it received as it was ground against bedrock in the sole of a moving glacier.*

A visit to any of the glaciers in the Park will reveal expanses of bedrock uncovered by retreat of the ice during the last few decades. These are good places to observe how moving ice erodes bedrock. Glaciated bedrock surfaces are usually very smooth, sometimes almost polished, and invariably display a pattern of parallel grooves rasped in the bedrock by other rocks dragged across it by moving ice. Such surfaces are littered with small rocks dropped from the melting ice. These, the tools that scoured the bedrock, usually have ground on them flat faces that are covered with crisscrossing patterns of small scratches.

Rocks held in moving glaciers are pulverized to a fine white powder by grinding against each other and against the bedrock of the valley walls and floor. This rock flour is released from the ice when it melts and is carried away by meltwater streams. Large quantities of rock flour color lakes and streams greenish white, making them look almost like milk. Lesser quantities color the water various shades of green and greenish blue. There are enough glaciers in Waterton-Glacier Park to supply small quantities of rock flour to many of the streams and lakes giving them their fascinating range of intense colors.

*Polished and grooved bedrock surface exposed by retreat of small glacier at foot of Mount Clements near Logan Pass.*

*Talus formation*

In places where melting ice left bare rock, erosion must begin by reducing the rock to soil. Under a northern climate, the first step in this process is taken when water seeps into cracks and freezes. The force of expansion exerted by water freezing within cracks breaks off fragments of rock which tumble to the base of the cliff where they accumulate to form a loose pile called talus, or scree. As frosts wedge more fragments loose from the cliffs and drop them onto the slopes below, the talus piles grow larger until ultimately they bury the last remnants of the cliffs that spawned them. So the bold cliffs left by the glaciers are being picked to pieces by the frost and slowly converted into smooth slopes of sliding angular debris.

*Soil formation*

Freezing water relentlessly attacks the rocks again and again, exploiting every minute fracture to break them finer. And every time a piece of rock is broken, new surfaces are exposed, thus hastening the process of disintegration.

*Glacial till east of lower St. Mary Lake sliding into the roadway. Extent of soil development is indicated by the thin, dark zone at the top of the exposure.*

*Beargrass and other plants shelter the soil of alpine meadows from erosion by running water. Danny On photo.*

Water on rocks also attacks all of the common silicate minerals, except for quartz, in another way by chemically changing them to varieties of clay. This process operates mostly on the surface of the rock so it is also hastened as new surfaces are exposed by frost splitting. The end result of attack by water on a silicate rock, such as the mudstones in Waterton-Glacier Park, is quartz grains embedded in clay; a sandy soil.

Limestones are composed of carbonate minerals which do not form clay when chemically attacked by water. Instead, they dissolve, making the water "hard," and are carried away in solution. Only the insoluble sand and clay impurities in the rock remain behind so limestones normally form very thin soils.

Of course fragments of rock contained in glacial deposits are attacked by soil-forming processes in the same way as the bedrock. Fragments of silicate rocks are reduced to clay and quartz grains while pieces of limestone dissolve away almost completely. Geologists use the extent of soil development on these deposits as a rough guide to their age. Material laid down during the last ice age usually shows very little soil development whereas deposits dating from earlier ice ages are deeply-weathered. Older glacial deposits exposed east of the Park are weathered to a distinctly reddish color and have lost all except the largest fragments of limestone. They contrast to deposits of the last ice age which are mostly light gray and contain numerous fragments of limestone still quite intact.

*Erosion of the Highline Trail near Logan Pass. Trampling feet of hikers have destroyed the protective vegetation exposing the soil beneath to impact of raindrops and causing water to run off the surface.*

Examining glacial deposits is a good way to gain an impression of how long it takes soil to form on material that has already been broken into small pieces. All the glacial deposits exposed in roadcuts within the Park boundaries date from the last ice age and have been exposed to weathering only since the ice melted no more than 10,000 years ago. In most places the soil that has formed on them in that length of time is only a few inches thick and still filled with stones. Evidently soil can not be regarded as a renewable resource within spans of time that have any meaning for human planning.

Plants are important because they determine what processes of soil erosion may operate. Their leaves shelter the soil from impacts of raindrops and their presence, along with that of the animal community they support, maintains an open, porous soil that encourages rainwater to soak into the ground instead of flowing across the surface. These combined effects greatly retard soil erosion in an area well covered by vegetation. Growing plants hoard the soil beneath them so that it grows deeper as the underlying bedrock disintegrates under the attack of the elements. Because plants control the kind of erosional processes that may operate, they control the form of the landscape these processes carve. The lush forests of Waterton-Glacier Park will dictate the eventual form of the post-glacial landscape now forming.

*A bear digging out a tasty marmot threw soil downslope. More will move down when the holes finally collapse and fill.*

*Soil and rocks torn up by the roots of a fallen tree move another step down the slope toward the stream that will eventually carry them off to the sea. Along "Trail of the Cedars" near Avalanche Campground.*

### *Soil creep*

Plants retard soil erosion but they cannot stop it entirely. Various slow processes combine to move the soil inexorably downslope, even beneath a dense forest.

Burrowing animals of all sizes from worms to bears dump the soil they excavate downhill as they dig holes that must eventually collapse from above, moving still more dirt downslope. Plant roots make way for themselves by shouldering aside the soil, presumably mostly in the downhill direction. Then when the plant dies and its root system decays, the holes left behind collapse from above moving more dirt downslope.

Soil expands when it is warmed, wetted, or frozen and shrinks when it is cooled, dried, or thawed. During every expansion the soil moves straight out from the side of the hill and then drops vertically downward when it contracts again. So every cycle of expansion and contraction hitches the entire soil mantle on a hillside a short distance downslope. The long term effect is as though all the soil on the slope were slowly pouring downhill.

Soil creep is the term applied to the downslope movement caused by burrowing animals, growing plants, expansion and contraction, and several other, less important, processes all working together. Because all these processes are more effective near the surface than at depth, the soil creeps fastest near the surface and the rate of motion decreases regularly with increasing depth.

*Reynolds Creek cascading down the glaciated floor of its valley. Danny On photo.*

Soil creep is too slow to be easily measureable but its effects are frequently obvious where deep soils or glacial tills lie on steep slopes. Posts set in such slopes are slowly pushed over by the more rapidly moving soil near the surface so that after a few years they lean downslope. Trees constantly strive upwards toward the light and develop trunks bowed downslope wherever rapidly-creeping soil tends to push them over.

Hillslopes dominated by soil creep are typically very smooth with rounded crests. Cliffs and other irregularities are rare because they are buried beneath the mantle of soil pouring slowly down the slope. Such landscapes exist today in parts of the northern Rockies too low to have caught enough snow to form glaciers during the ice ages.

Steeply-sloping glacial deposits and areas of the eastern part of the Park where Cretaceous rocks outcrop are the only places in Waterton-Glacier Park where soil creep is an important process today. Frost splitting and talus formation are still the dominant erosional processes on most of the rockier slopes in the Park. But as time passes and thick mantles of soil slowly form on the talus slopes, the domain of soil creep will grow larger and eventually extend over most of the Park unless another radical change in climate intervenes.

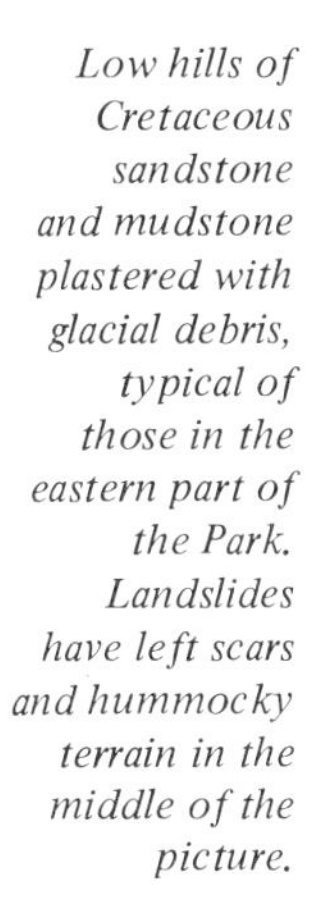

*Low hills of Cretaceous sandstone and mudstone plastered with glacial debris, typical of those in the eastern part of the Park. Landslides have left scars and hummocky terrain in the middle of the picture.*

## *Landsliding*

If a hill is steepened to an angle greater than the soil can hold, it will eventually slide. Excess water may weaken the soil by saturating it and thus cause sliding. Waterton-Glacier Park contains numerous landslides that owe their origin to various natural and manmade combinations of these two causes.

*Soil creep has given the trunks of these young trees growing beside McDonald Lake their downhill slouch.*

Landslides move in a distinctive way. A fracture surface curved like the bowl of a spoon develops in the slope and all the material above it moves down as a mass while that beneath remains in place. Rate of movement is occasionally quite rapid but more often very slow, perhaps amounting to no more than a few inches per year. Slides leave a hollow scar high on the slope where the movement started and an irregularly hummocky deposit below where the slide mass stopped. Wherever hummocky patches beneath small, curving cliffs mark the hillslopes, we can be sure that landsliding has been an active erosional process. The slide across from the "loop" on the Going-to-the-Sun Road, pictured on page 36 is a good example.

In most areas landslides are unusual because the hills long ago adjusted their slopes to stable angles. Landslides tend to be more numerous in populated areas where construction has steepened slopes and interfered with drainage. Numerous small landslides along many of the roads in Waterton-Glacier Park are the direct result of such human interference. Other landslides in the Park are completely natural, the result of glacial steepening of hillslopes and interference with drainage.

Landslides are especially common east of the Lewis Overthrust Fault where the hills are underlain by very weak rocks: Cretaceous sandstones and mudstones heavily plastered with glacial deposits. Numerous layers of clay shale within these deposits become very soft and slippery when they are soaked with water. Glacial erosion and deposition compounded the problem by disrupting the natural drainage to create ponds and marshes from which water soaks into the ground weakening the shale layers and causing slides. Many of the glacial deposits were left on very steep slopes from which they slide quite readily, especially after heavy rains.

### *Mudflows*

As the name suggests, mudflows are simply thick slurries of dirt and water. They may develop from landslides if these involve very wet soil or they may develop when water flowing in a stream picks up an unusual load of sediment. In either case they move a large quantity of material in a very short period of time.

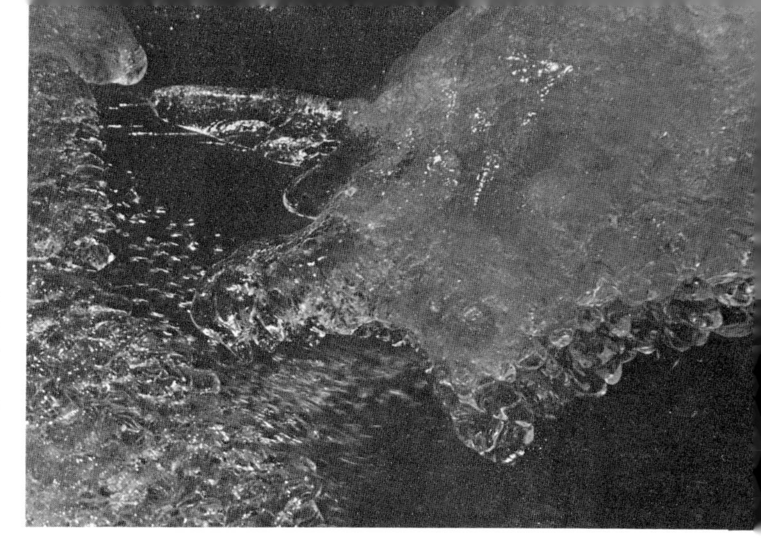

*Spring's melting ice freshens the streams. Jim Mohler photo.*

Everyone has noticed that it is easier to lift a rock under water than in the air because the submerged rock is buoyed up by the weight of the water it displaces. Rocks submerged in mud are even more strongly buoyed up because of the greater weight of the displaced volume. So mudflows are capable of transporting surprisingly large boulders. Torrential rains that precipitated the disastrous floods of June, 1964 also caused mudflows in Waterton-Glacier Park. Remnants of some of these are still apparent.

*Rockfalls*

Bedrock is never really solid. It always contains fractures or other weak surfaces that may break. This is especially likely to happen in a place like Waterton-Glacier Park where numerous very steep cliffs were left unsupported when the ice melted at the end of the last glacial period.

Small rockfalls are common and usually do nothing more than dump a large quantity of broken rock at the base of a cliff, leaving a corresponding scar above. An excellent example is visible at the south side of St. Mary Lake near the narrows where a fresh scar left by a recent rockfall marks the Altyn Limestone. A large rockfall slightly altered the profile of Chief Mountain during the summer of 1972.

Very large rockfalls occasionally trap a cushion of air beneath them in the same way that avalanches do and move very rapidly and destructively. This has not happened in the Park, at least not recently, but such rockfalls have occurred in nearby areas.

*Winter in Red Rock Canyon. Danny On photo.*

## *Avalanches*

During the warming days of late winter and spring large cornices of snow break off the high slopes and come racing down the mountainsides as avalanches. These frequently move very rapidly because the falling mass of snow traps a cushion of compressed air beneath itself and then rides it down the mountain. Avalanches frequently bring trees with them and sometimes rocks and dirt as well, so they can contribute to the erosion of the mountainsides. They tend to occur repeatedly in the same places because the prevailing winds build a new cornice of snow in the same place each winter. Frequent repetition of avalanches maintains cleared tracks often called "green slides" because they make a green path down the mountain.

*Streams*

Streams are a special joy in Waterton-Glacier Park. They bubble brightly over beds of colored pebbles of Precambrian mudstone as they resume a major role in shaping the landscape now that the glaciers have almost entirely melted. They function mostly as the conveyor belts of the landscape removing sediment brought down to them by the various processes of erosion operating to shape the hillslopes.

Heavy vegetation that protects the soils in the Park from rapid erosion during heavy rains also dictates the character of its streams. Forested soils are open and porous and behave as though they were a giant sponge soaking up water when it is abundant and then letting it seep out again during dry seasons. They prevent streams from flooding during all but the wettest seasons and from drying up during dry seasons.

Most of the time the streams in Waterton-Glacier Park carry remarkably clear water, obviously without much sediment. Like most streams, they carry their sediment load during floods, especially during the occasional very heavy floods that may come only once or twice in a century. During June of 1964 a prolonged downpour of warm rain fell on a deep snowpack releasing raging floods that scoured streambeds moving even the largest boulders and leaving scars that will remain obvious for decades. Most of the sediment will go no farther until the next catastrophic flood.

*Debris-strewn floodplain of Cameron Creek as it looked shortly after the major flood of June, 1964. Canada Parks photo.*

*The area under the bridge approximately defines the gorge McDonald Creek has cut into solid bedrock since the glaciers melted. The creek has been working its way down the slope of tilted layers in the bedrock making the gorge triangular in cross-section. Near McDonald Falls.*

Streams in Waterton-Glacier Park have cut their channels deeper since the big glaciers melted eroding small gorges where they flow on bedrock. As they erode their channels deeper, the streams also cut them to a flatter slope thus slowing the velocity of the water. Eventually the streams will reduce their beds to a slope gradient gentle enough to stop the downcutting. Regardless whether streams are flattening the gradient of their beds by eroding or steepening them by depositing, they always act to create a slope exactly steep enough to enable the water to flow without either eroding or depositing. Any change in watershed conditions that causes variation in supply of either water or sediment to a stream is likely to cause it to begin eroding or building up its channel. Forest fires, severe logging or overgrazing, for example, may increase the rate of erosion in the watershed thus increasing the amount of sediment the stream must carry and causing it to build its bed up to a steeper gradient by depositing sediment. Streams respond quickly and sensitively to any interference with their flow or change in condition of their watersheds.

*Baring Falls, downstream from Sunrift Gorge, one of the numerous waterfalls in Waterton-Glacier Park. Streams sandblast falls away unless their water is clear.*

*Solid rocks sculptured and polished by sediment carried in rushing stream water. Cylindrical potholes are drilled by pebbles turning on the bottom in the grip of these whirlpools on McDonald Creek.*

Velocity of stream flow depends not only upon the steepness of the bed but also upon its roughness. Large boulders and other objects tend to cause friction between the flowing water and the walls of its channel thus slowing stream flow. Any kind of human interference with the roughness of the channel will affect the velocity of the stream. Smoothing the stream bed, by removing boulders for example, will have the effect of increasing flow velocity and also increasing hazard of flooding at points downstream. Deepening and straightening a stream channel will likewise increase flow velocity and hazard of flooding downstream.

Streams are capable of eroding solid rock only when the water is loaded with sediment and can sandblast its bed to carve rounded forms with smoothly-polished surfaces that suggest abstract sculpture. Persistent whirlpools help by seizing pebbles in their grip and swirling them around, slowly grinding cylindrical potholes into solid rock. Look for these in late summer when the water is low.

Streams tend naturally to follow a meandering course. They find it difficult to do this in bedrock because there they are not free to follow a course of their own choosing. And streams heavily loaded with sediment, such as those draining directly from glaciers, also have difficulty meandering because they constantly choke their own channel with deposits of debris. These pick their way, in a crisscrossing braided pattern, through beds of gravel and sand. But most streams flowing across easily-erodible beds of material such as glacial till or outwash generally follow a freely-meandering course.

Meander bends are never stable because streams erode material from their outside banks, where the current strikes and the channel is deepest, and deposit on the inside banks where the current is slack. So the stream persistently tends to shift its channel toward the outsides of meanders making them grow larger. However, their growth is limited because the stream eventually cuts across the neck of the growing meander leaving it abandoned as a shallow, horseshoe-shaped pond that slowly fills with silt to become a swamp and finally disappears. Constant shifting of meander bends enables the stream to work its channel back and forth across the entire flood plain.

*Trees toppled by McDonald Creek undercutting their root systems as it erodes the outside of a meander bend.*

*Abandoned meander of McDonald Creek left as a placid pond when the stream shifted its course.*

Freely flowing and undisturbed wild streams arrange their beds in a series of deep pools, at the outsides of bends, and shallow riffles, between bends. Spacing of these pools and riffles, like the size and spacing of meander bends, is directly related to the width of the stream. So it is perfectly normal to see streams winding along their valley floors through a regular succession of predictably spaced pools and riffles.

### *Lakes*

When the great ice age glaciers melted, they left the landscape of Waterton-Glacier Park liberally gifted with lakes of all sizes. Some of these, especially the smaller ones in the high valleys, occupy basins scooped out of hard bedrock. Others, mostly the larger ones low in the valleys, are impounded by dams of glacial till or outwash. Still others, especially the small prairie ponds, are simply holes left when blocks of ice melted after being buried in glacial deposits.

Whatever its origin, a lake is always destined to be a temporary feature of the landscape. Processes of erosion now active in the Park will eventually conspire to destroy those left by the ice. No process operates to maintain lakes.

*Waves breaking on the shore have created a beach and undercut the shoreline of McDonald Lake. Rapid soil creep causes the tree trunks to bend downhill.*

Streams entering lakes bring sediment, especially during floods, and it all stays in the basins, gradually filling them. Fortunately for the lakes in the Park, this will be a slow process because the streams carry very little sediment. Nevertheless, most of the major lakes already have sizeable deltas built into them at the mouths of the larger streams. Outlet streams tend to erode their spillway channel deeper thus lowering the lake level and eventually draining it. This will also be a slow process because the clear streams draining lakes enclosed in bedrock basins carry very little sediment to sandblast their channels. Even streams draining lakes impounded behind deposits of glacial debris may have a hard time; many of these deposits are composed of very coarse pebbles and boulders too large for the modern outlet streams to move.

But none of these fortunate circumstances will be enough to finally save the lakes of Waterton-Glacier Park. Ultimately, thousands of years from now, they will all be filled and drained and the streams will run continuously down their valleys. When that day comes, the landscape of the Park will depend for standing water upon meander loops left abandoned by streams as they slowly shift their channels back and forth across their floodplains.

*Frozen springs add their own touch to the beauty of the winter. Jim Mohler photo.*

*Pioneering plants struggle to cover the Garden Wall.*

# Roadguides

## GOING-TO-THE-SUN-ROAD

### *West Glacier – Avalanche Campground*

West Glacier and Apgar lie at opposite ends of a short valley between the Apgar Mountains to the northwest and the Belton Hills to the southeast. Too large to have been eroded by McDonald Creek, which now flows through it, this valley must have belonged originally to the North Fork of the Flathead River until large glaciers diverted that stream into its present course north of the Apgar Mountains. Now the valley is floored by thick deposits of glacial outwash well exposed at West Glacier where the Middle Fork of the Flathead River has cut a gorge through them.

A very small, rounded hill rises east of the road immediately south of the Park entrance station between West Glacier and Apgar. Material exposed in the roadcut is glacial till, so evidently the hill is part of a moraine now nearly buried in outwash sand and gravel swept in by meltwaters from the glacier that scooped out the basin of McDonald Lake. The buried moraine and outwash deposits together make the dam that impounds the lake.

*The Garden Wall as seen from Logan Pass looking north toward Granite Park.*

Both the Apgar Mountains and the Belton Hills are parts of the Whitefish Range, another great block of Precambrian sedimentary rocks similar to those in the Park. Eastward dipping strata exposed along the road where it passes close to the base of the Belton Hills, along the lower end of McDonald Lake, belong to the Siyeh and Shepard Formations.

*Avalanche Creek has cut a small gorge into red mudstones of the Grinnell Formation. Danny On photo.*

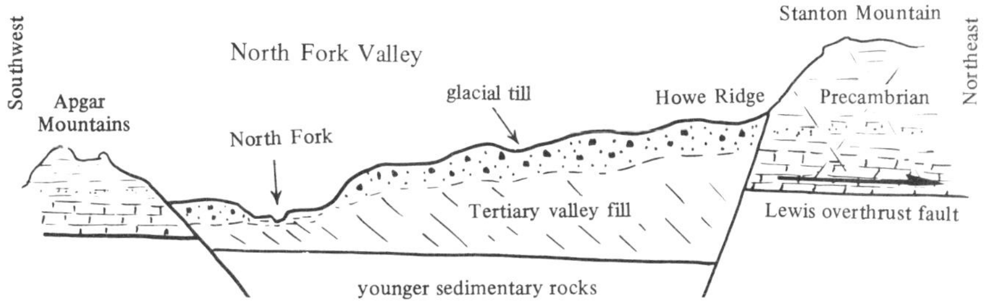


*Southwest-northeast cross-section through Howe Ridge along the north side of McDonald Lake.*

Between Apgar and McDonald Falls the road winds for about ten miles through dense forests growing on glacial till along the south shore of McDonald Lake. The lake fills a basin scooped out of soft deposits of sand and gravel that accumulated in the North Fork Valley during the Tertiary Period. Because of the thick deposits of glacial debris, there are no good exposures of Tertiary valley-fill sediment along the road. The only outcrops are a few beds of sticky clay exposed in the beaches near the northeast end of the lake during late summer when the water level is low.

The North Fork Valley first began to form when the mass of Precambrian sedimentary rocks in the Park moved eastward on the Lewis Overthrust Fault pulling away from the Whitefish Range to the west. Then the floor of the valley appears to have dropped and, finally, it received its filling of sand and gravel. Strata in the mountains on opposite sides of the North Fork Valley were continuous before the valley opened up behind the eastward-moving Lewis Overthrust slab.

Between the head of McDonald Lake and Avalanche Campground, McDonald Creek flows alternately across hard Precambrian bedrock and loose deposits of glacial debris left behind as the ice melted back up the valley at the end of the last ice age. Waterfalls and rapids are common in the bedrock reaches where the stream is trapped in narrow gorges.

Where McDonald Creek flows on an easily-erodible bed of glacial outwash gravels, and is not trapped in a channel by bedrock walls, it follows a freely-meandering course. Old meanders and abandoned lengths of stream channel are cut off and left behind as ponds while the stream works its way back and forth across its floodplain. Long, narrow ponds south of the road in "Moose Country" are typical segments of abandoned stream channel. Eventually these will fill with sediment brought in during spring floods of muddy water. But the stream will continue to produce new ones as the old ones fill, so its floodplain will never lack summer habitat for moose and other water loving animals.

Greenish mudstones of the Appekunny Formation outcrop along the lower course of McDonald Creek to within about a mile below Avalanche Campground where alternating green and red layers signal the transition to the red mudstones of the overlying Grinnell Formation.

### *Avalanche Campground – Logan Pass*

Avalanche Campground is at the mouth of Avalanche Creek which has its headwaters in Sperry Glacier. Glacially-pulverized rock flour suspended in the water gives it various hues of green depending upon how much is being released by melting ice above. Avalanche Lake occupies a large basin about two miles up the valley, an easy hike.

A short nature trail, the "Trail of the Cedars," winds through a dense forest of western red cedars growing along the north banks of Avalanche Creek near the campground. The creek flows through a narrow gorge it has cut into red mudstones of the Grinnell Formation during post-glacial time, an excellent place to see stream-sculptured rock.

For several miles upstream from Avalanche Campground the road follows the floodplain of McDonald Creek which flows in places on bedrock but mostly over a bed of glacial gravels brightly colorful with pebbles of red and green mudstone. No doubt the stream would by now have cut a deep gorge through the easily-erodible gravels if it did not have to erode through bedrock as well. The rate at which the stream is able to carve through the most resistant rocks in its course determines how fast it may deepen its valley.

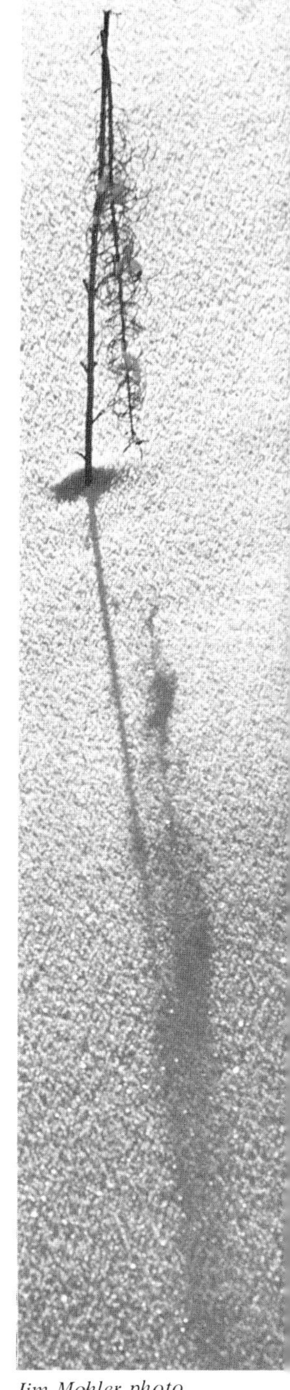

*Jim Mohler photo*

*Stromatolite reef in the Siyeh Formation. Exposed in roadcut beside Going-to-the-Sun Road above the "loop" west of Logan Pass. Plants indicate the scale.*

Forest fires burned both sides of the valley of McDonald Creek across the line of the Going-to-the-Sun Road during August, 1967. Vigorous second growth of bushes and small plants has effectively protected the soil from rainsplash erosion even though the original forest was reduced to blackened snags.

All along the grade between McDonald Creek and the summit of Logan Pass, there are outcrops of Siyeh Limestone. Nearly every bed contains fossil algae, their shapes preserved in the crusts of calcium carbonate that formed over a billion years ago on scummy mats of algae growing where the sun shone through shallow water. Algae grew so abundantly that they left the entire formation stained by organic remains of their tissues; look at the dark color of freshly-broken surfaces. On older surfaces algae are etched into low relief by weathering.

A large landslide scar (illustrated on page 36) is conspicuous on the valley wall above McDonald Creek directly across from the "loop" in the Going-to-the-Sun Road. Evidently the ice plastered lateral moraine onto bedrock valley walls so steep that the glacial till couldn't hold the slope and slid down into the valley of McDonald Creek. Most of the lateral moraine is still where the ice left it and easy to recognize because the glacial till supports a lush forest that ends abruptly along a horizontal line high on the valley wall. Heavy forests growing below this line define the former extent of the ice.

*View of Going-to-the-Sun Road above McDonald Creek showing cracks caused by slow landsliding.*

Flattop Mountain, visible farther north up the valley of McDonald Creek, is most descriptively named. Its summit is an extensive, gently-rolling plateau quite unlike the jagged spires that form most of the nearby peaks. Apparently it was completely covered by an ice cap, the source of the mighty glaciers that poured southward down McDonald Creek and northward down Waterton Creek. The entire top of Flattop Mountain was planed smooth by the ice.

*McDonald Creek creates a gorge by cutting back the lips of waterfalls. Tom McBride photo.*

*The vertical band of black rock beside the upper portal of the tunnel is a dike of diabase cutting across nearly-horizontal layers in the Siyeh Limestone.*

Travellers ascending the long grade up the Garden Wall far above the valley of McDonald Creek notice numerous cracks and semi-circular patches in the outer lane of the road. These mark places where small landslides are slowly moving segments of the road down the slope presenting an interesting problem in maintenance. This is an excellent example of landsliding caused by human interference with a hillslope. Numerous culverts and drains have been installed in an attempt to keep the ground beneath the road dry enough to prevent sliding.

### *Logan Pass – Sunrift Gorge*

Ice flowed both ways from Logan Pass, eroding a deep notch in the Continental Divide and creating a magnificent alpine landscape. Effects of glacial erosion and deposition are still perfectly fresh at this altitude where winter is never more than a few weeks away, even summer weather frequently makes ice-age conditions easy to imagine.

Logan Pass is in the top of the Siyeh Limestone where it passes transitionally into the mudstones and limestones of the overlying Shepard Formation. Red mudstones of the Kintla Formation conspicuously cap some of the surrounding mountains. Just beneath the level of the pass, and exposed along the road on both sides of it, is the big Purcell Sill, the 100-foot-thick layer of black diabase injected as molten magma between beds of the Siyeh Limestone.

The visitor center at the pass is built on a thick ledge of Siyeh Limestone containing numerous large stromatolite heads, a form of blue-green algae believed to have grown where waves agitated the water. Several other ledges containing beautiful stromatolites of all sizes outcrop along the trail between the visitor center and Hidden Lake. Layers containing stromatolites are slightly more resistant to weathering and erosion than the others and usually outcrop as prominent ledges. Mudstones of the Shepard Formation, exposed along the higher parts of the trail to Hidden Lake, contain perfectly preserved fossil mudcracks, ripple marks, and numerous other sedimentary features. A hike along this trail is a walk through a natural museum filled with exhibits from Precambrian time.

Mount Clements, which rises north of the trail to Hidden Lake immediately west of the visitor center, has a snowpatch at the eastern side of its base which was a small glacier until very recently. Very little of the snow or ice is visible from the trail or the visitor center because it is hidden behind a very young moraine, heaps of broken rubble looking as raw as though a bulldozer had just piled them there. Wonderful exposures of glacially polished and scratched bedrock surfaces recently revealed by melting ice are reached by hiking behind the moraine.

*Water seeping between layers of Siyeh Limestone beside the Going-to-the-Sun Road.*

*Exposure of the top of the Purcell Sill in a roadcut on the east side of Logan Pass. Black diabase cuts irregularly across lighter-colored layers of limestone.*

Highline Trail follows the west face of the Garden Wall between Logan Pass and Granite Park, about 6 miles north, where there are good exposures of the Purcell lava flows. Just a few hundred yards north of Logan Pass, where it traverses its first cliff face, the Highline Trail crosses an excellent exposure of the Purcell Sill. The fresh diabase is a very dark gray rock, almost black, composed of white crystals of plagioclase about the size of rice grains, scattered through a matrix of black pyroxene. Water circulating along fractures has coated them with green minerals that often form when water attacks diabase. Just above the sill is a zone of white marble formed where organic matter was driven out of the Siyeh Limestone by heat from the molten magma. Fossil algae are especially easy to see in the bleached limestone.

Between Logan Pass and Siyeh Creek the Going-to-the-Sun Road traverses the southern and eastern faces of Piegan Mountain, passing exposures of Siyeh Limestone. Reynolds Creek, a tributary of the St. Mary River, works its way along the rugged floor of its glacially-gouged valley almost 1500 feet below the road.

Most of the road between Siyeh Creek and the head of St. Mary Lake is built on a lateral moraine plastered onto the north valley wall of Reynolds Creek. Glacial till is exposed in the roadcuts. It is common to find lateral moraines best developed on the north side of a valley because this is the side that receives the most sunshine and experiences more temperature changes. During the ice age more loose rock was dropped onto the ice from the north valley wall by freezing and thawing than from the opposite wall which remained frozen more of the time.

Bedrock outcrops between Siyeh Creek and Sunrift Gorge are all in the Grinnell Formation, mostly red mudstones with occasional layers of green mudstone or white sandstone. All of the exposures are full of interesting sedimentary features such as mudcracks, ripple marks, and raindrop impressions, but outcrops containing beds of white sandstone are the best ones to examine.

*Mudchip breccia, small pieces of red mudstone mixed into white sandstone, in the Grinnell Formation. Pieces of sun-dried mud were picked up by running water and mixed with sand to form this rock.*

*Intense heat from the molten magma that was intruded to form the black diabase of the Purcell Sill bleached and recrystallized the Siyeh Limestone to form white marble near the contact. The thin black band within the marble is a small offshoot of the sill. Highline trail near Logan Pass.*

Outcrops around Sunrift Gorge, near the head of St. Mary Lake, show alternating layers of red and green mudstone. They are in the transition zone between the Grinnell Formation and the green mudstones of the Appekunny Formation beneath.

*Glaciers carved deep cirques into these mountains leaving jagged spires. Marion Lacy photo.*

*Baring Creek flowing through Sunrift Gorge, a crack that opened when the rock on the left slid a few feet downhill. This is a highly unusual way for a canyon to form.*

### *Sunrift Gorge — St. Mary*

At Sunrift Gorge, Baring Creek flows through a deep crack for several hundred feet following a perfectly straight course. Careful comparison of Sunrift Gorge with the much shallower and different looking gorge Baring Creek has eroded elsewhere in its course, makes it clear that the Creek did not erode this cleft. Apparently a large block of Precambrian mudstone slid a few feet downslope on a bedding surface, opening the crack behind it.

Someone who cared about rocks built the wall around the parking area at Sunrift Gorge. It contains excellent specimens of nearly every important rock type in the Park even including the bleached Siyeh Limestone next to the Purcell Sill; an excellent outdoor museum where visitors can try to figure out how the specimens should be labelled.

Between Sunrift Gorge and Rising Sun Campground, the Going-to-the-Sun Road winds along the north shore of St. Mary Lake past large outcrops of green Appekunny mudstones and, near the narrows in the lake, buff-colored Altyn Limestone. Immediately beneath the Altyn Limestone is the Lewis Overthrust fault and beneath that the much younger Cretaceous sedimentary rocks.

From the narrows and Rising Sun Campground east, the bedrock is Cretaceous sandstones and mudstones heavily-plastered with glacial debris. These rocks form rounded hills very different in appearance from the bold mountains eroded in the much harder and stronger Precambrian sedimentary rocks above the fault.

The hummocky, forested ridge that rises about 500 feet above the south side of St. Mary Lake is a large lateral moraine that extends almost to the community of St. Mary. However, the ridge no longer extends across the lower end of the lake as a terminal moraine; evidently it was eroded away by outwash streams. So St. Mary Lake is not impounded by a moraine. Instead, it appears to be dammed behind large deposits of glacial outwash swept into the St. Mary valley from tributary creeks on both sides of the valley. (See diagram, page 99).

Singleshot Mountain, a long ridge with a flat top, rises prominently on the skyline immediately north of St. Mary. Its upper part is in the Appekunny Formation which contains thick beds of white quartzite easily visible from a distance. Below the Appekunny is a thin, buff-colored zone, the Altyn Formation, which rests directly on the Lewis Overthrust Fault.

*View of distant peaks at the head of the glacially straightened trough of St. Mary valley. Marion Lacy photo.*

*Baring Falls pours over layers of red and green mudstones in the transition zone between the Grinnell and Appekunny Formations. Danny On photo.*

*A cliff of Altyn Limestone rises like a wall west of Rising Sun Campground marking the place where the road crosses the Lewis Overthrust Fault.*

*Trick Falls in late summer with most of the flow issuing from the cavern and only a trickle coming over the lip.*

## TWO MEDICINE ROAD

### *U.S. 89 – Two Medicine Lake*

Between U.S. 89 and Two Medicine Lake the road winds for almost 8 miles through a spectacular glacially-gouged trough. East of where the road crosses Two Medicine Creek it is built on Cretaceous rocks, now plastered with glacial debris, that were exposed as the creek eroded down through the Precambrian rocks above the Lewis Overthrust. West of the creek crossing, the road is built mostly on Precambrian rocks above the Lewis Overthrust Fault.

Lower Two Medicine Lake is a reservoir impounded by a dam built for irrigation purposes before the Park was established. The dam was severely threatened during the flood of June, 1964 when large quantities of sediment were washed into the reservoir building several small islands still visible just above the dam.

Trick Falls is at the end of a beautiful trail through the woods about a quarter of a mile from the road. The falls are where a small stream pours over a cliff in hard Altyn Limestone, above the Lewis Overthrust, onto soft Cretaceous mudstones beneath. Altyn Limestone, like most limestones, occasionally develops caverns where percolating ground water has dissolved the water-soluble rock. Apparently part of the water enters the falls through a small cavern, about half way up the cliff, while the rest comes over the lip in the normal fashion. Early in the summer when water is plentiful Trick Falls looks like any other waterfall. But during dry seasons most of the water comes out of the cavern and the surface stream pouring over the top of the falls nearly dries up. At this time of year the trickle of water still flowing over the top of the falls all comes from a small tributary that meets the main stream just above.

Two Medicine Lake occupies a basin scooped out of Precambrian bedrock where three glaciers met to form a mighty river of ice. The sharp prow of Sinopah Mountain, glacially-carved from red mudstones of the Grinnell Formation, cuts the skyline west of the lake. The lower part of this mountain is still relatively smooth where it was scrubbed clean by the ice. The upper part, not buried in ice during the last glacial period, is very rough and shaggy-looking, the effects of frosts splitting the rock during the last ice age. Look for the line separating smooth slopes below from rough slopes above to get an impression of the depth of ice that filled this valley during the last ice age.

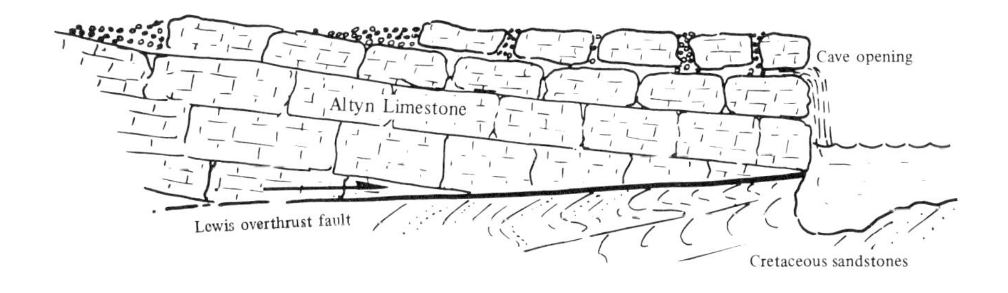


*Trick Falls drawn along the stream showing how water soaks into cavernous limestone below the stream bed. The Lewis Overthrust fault is directly beneath the Falls.*

Iceberg Lake.
Danny On photo.

## MANY GLACIER ROAD

### *U.S. 89 – Swiftcurrent Lake*

The Many Glacier Road turns west from U.S. 89 at Babb and follows the valley of Swiftcurrent Creek 12 miles into the mountains.

Swiftcurrent Creek has eroded through the slab of Precambrian rocks above the Lewis Overthrust into the Cretaceous sandstones and mudstones beneath. Most of the Many Glacier Road is built on Cretaceous rocks in the valley floor between valley walls cut in much older Precambrian rocks. The road finally crosses the Lewis Overthrust onto the Altyn Limestone about one mile east of Swiftcurrent Lake.

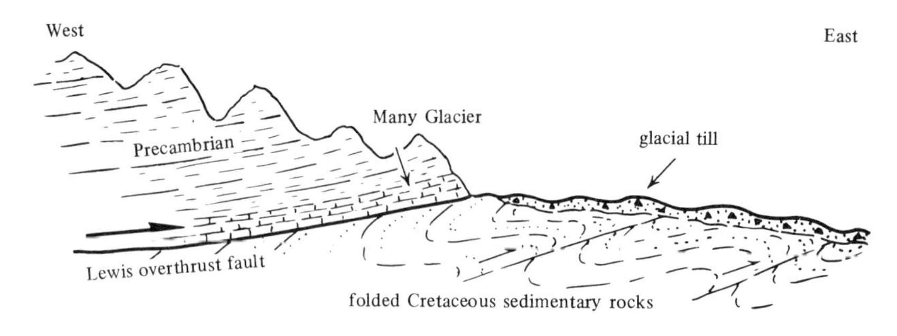


*West-east cross-section along the line of the Swiftcurrent Valley.*

*View looking across Sherburne Reservoir towards Grinnell Glacier barely visible as a white patch in the right center.*

Sherburne Reservoir is impounded by a dam built in 1921 as part of an elaborate irrigation scheme inaugurated before the Park was established. Water from the reservoir goes over the Hudson Bay Divide, to the east, into the Milk River drainage basin where it is used for irrigation. The reservoir floods the site of the first oil field in Montana where small quantities of petroleum were produced from shallow wells during the early years of the century.

Outcrops of Cretaceous rocks are hard to find along the Many Glacier Road because the valley floor is nearly covered with glacial till. But there are a few roadcuts, especially beside Sherburne Reservoir, that expose small amounts of black mudstone. This appears to be the material responsible for the series of small landslides that slowly carry the road toward the Reservoir, making annual repairs necessary.

Large cliffs of tan Altyn Limestone beside the road mark the spot, within sight of the Many Glacier Hotel, where it crosses the Lewis Overthrust onto Precambrian rocks. The entrance to one of the mines opened during the copper boom that flared briefly in this area at the turn of the century is visible in a large cliff of Altyn Limestone north of the road. Look for a small hole in the cliff with a large debris-covered slope beneath it.

Swiftcurrent Lake is scooped out of solid Precambrian rock in a place where the layers tilt downward to the west, making it easy for the glacier to gouge a hollow by peeling off the strata. The outlet stream is slowly draining the lake by cutting down the lip of the small waterfall on the entrance road to the Many Glacier Hotel.

The view westward from Swiftcurrent Lake toward the head of the valley in a series of cirques along the east side of the Garden Wall is as geologically interesting as it is scenically magnificent. Grinnell Glacier nestles low in the scene deep in the shadow of the Garden Wall. Above it, in the gray limestone cliffs of the Garden Wall, the Purcell Sill is continuously exposed for several miles as a black ribbon of diabase bordered above and below by white bleached zones in the Siyeh Limestone. Several small notches in the skyline at the heads of snow-filled chutes mark exposures of the diabase dikes that brought small amounts of copper minerals into these limestone cliffs.

*The North Fork of the Flathead River has cut through gravelly glacial deposits to expose fine-grained sands of the Tertiary valley fill. Springs visible in the stream bank are ground water seeping from the base of the gravel.*

*The low, tree-covered ridge across Bowman Lake is a lateral moraine left by the glacier that scooped out the lake basin. A terminal moraine impounds the lake.*

## NORTH FORK ROAD

### *Apgar – Kintla Lake*

Most of the 40 miles of road between Apgar and Kintla Lake are narrow and unimproved so few park visitors pass this way. Dense forest fills most of the North Fork Valley reducing scenery along the road to an occasional glimpse. But this road leads to several of the finest lakes in the Park splendidly set in deep mountain valleys.

The North Fork Valley appears to be a gap that opened behind the slab of Precambrian rocks sliding eastward on the Lewis Overthrust. The valley floor probably dropped downward after that slab moved. Whatever its origin, it accumulated thousands of feet of valley-fill sediment while the region had an arid climate. Then this accumulation was partially removed by the North Fork of the Flathead River and its remnants covered by glacial till during the last ice age.

Now the North Fork Valley is floored by low, rolling hills that support a somber forest of conifers. Rainfall is relatively heavy in this western side of the Park.

*Fossil algae, in this case a pink stromatolite, often make colorful patterns in the rocks. Danny On photo.*

Most of the outcrops along the road are glacial till. But there are a few exposures of the valley-fill, recognizeable as soft-looking, nearly-white sands deposited in layers that now tilt downward to the east. Although fossils of vertebrate animals have been found in these beds, they are rare and not likely to be seen without a diligent and time-consuming search. Coal beds are visible where they were once mined, opposite the mouth of Logging Creek about 20 miles north of Apgar.

The series of large lakes along the east side of the North Fork Valley all occupy basins scooped, at least in part, from the soft valley-fill sediments. Some of these lakes, most obviously Bowman and Kintla Lakes, are dammed by moraines.

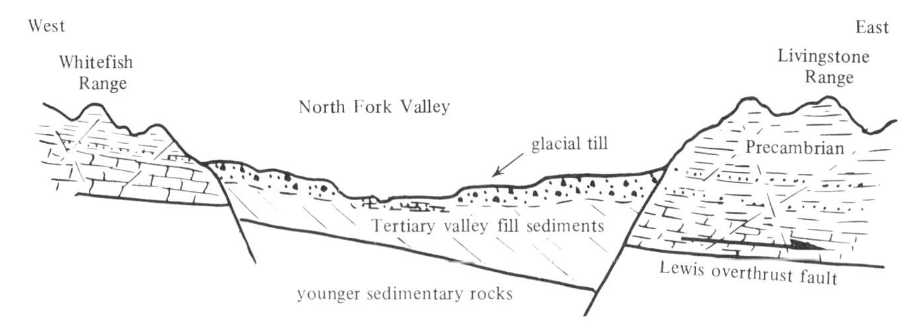


*West-east cross-section across the North Fork Valley.*

*Chief Mountain, an outpost of Precambrian rock isolated from the main mass by erosion. Tom McBride photo.*

## CHIEF MOUNTAIN INTERNATIONAL HIGHWAY

### *U.S. 89 (Babb) – Canadian Border*

The Chief Mountain Highway traverses till-plastered Cretaceous rocks of the high plains for the entire 15 mile distance between U.S. 89 and the entrance to the Canadian portion of the Park. Mountains of the Park rise abruptly above the rolling hills of the plains along the line of the Lewis Overthrust to the west.

Chief Mountain is unmistakeable, a tower of Precambrian rock standing alone, an outpost in front of the main mass of the mountains. It is a remnant isolated by erosion from the rest of the Precambrian slab that slid east on the Lewis Overthrust so that now it stands surrounded, like an island, by Cretaceous rocks.

*View of Chief Mountain as it appears from the east.*

Chief Mountain is the easternmost portion of the Precambrian rock that moved on the Lewis Overthrust. It is important to geologists because its position helps establish a minimum figure for the total eastward movement of the overthrust slab. Of course, no one can be quite sure how much Precambrian rock once east of Chief Mountain may have been entirely removed by erosion.

***Waterton Park Overlook – Waterton Park Entrance***

During the last ice age Waterton Park overlook would have been near the edge of a sea of ice that filled the valley below leaving only the upper halves of the peaks exposed. Glacial ice poured down the mountain valleys onto the plains where it gathered into a large ice field that spread about 10 miles northeastward. There it met the continental ice sheet flowing southwestward from the area of Hudson Bay, more than 1000 miles away (see map on page 29). When all this ice melted, approximately 10,000 years ago, the area once buried beneath glaciers was left covered by a hummocky mantle of till and outwash. The big roadcut of till across the highway from the overlook was dumped at this time.

Rugged mountains of Waterton Park, carved from the hard Precambrian sedimentary rocks, rise abruptly above the line of the Lewis Overthrust north of the overlook. When the light is right, huge folds are visible in the face of Mount Crandell showing how the layers in the Precambrian sedimentary rocks were rumpled as they dragged across the Cretaceous sedimentary rocks beneath.

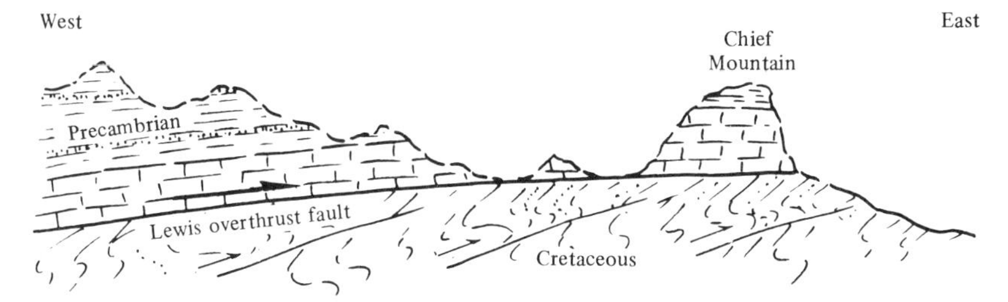


*West-east cross-section through Chief Mountain, a rootless monolith of Precambrian rock separated by erosion from the rest of the mountains.*

Most of the road between the overlook and Waterton Park is built across glacial till dumped by the Waterton Valley glacier. Just outside the Park entrance, the road crosses the smooth surface of glacial outwash gravels near the outlet of Maskinonge Lake.

*Prince of Wales Hotel is perched on a low ridge that looks like a moraine but is actually a ledge of Altyn-Waterton Limestone thinly plastered with glacial debris.*

*Cameron Falls in the winter. Danny On photo.*

## WATERTON PARK ROADS

### *Waterton Park Entrance – Waterton Townsite*

The road follows a lateral moraine deposited by the big Waterton Valley glacier. Flat areas on the valley floor are underlain by outwash gravels swept in by torrents of meltwater released while the glacier was melting. Lower Waterton Lake fills a shallow depression created where a large patch of ice was left behind and partially buried in outwash gravels as the main glacier melted back. When the ice finally melted the lake basin was left where it has been buried.

Mount Crandell and Vimy Peak, on the opposite side of Waterton Lake, are both composed of tan Altyn-Waterton Limestone overlain by greenish-gray Appekunny mudstones. Folds and faults in the limestone are visible in the lower parts of both mountains.

*View looking north from above Logan Pass towards Waterton Park. National Park Service photo.*

Waterton Townsite is built on the gently-sloping surface of a fan-shaped deposit of outwash gravels swept into the Waterton Valley from Cameron Creek. Waves washing onto the shore have created an attractive beach in this material which slopes just steeply enough to enable waves to work the gravel back and forth as they break.

*The big glacier that scoured Waterton Valley undercut the mouth of Cameron Creek so that it is now a hanging valley. Water pours over a ledge of Altyn (Waterton) Limestone. Cameron Falls.*

*Hummocky moraine and outwash gravels near the Buffalo Paddock on Alberta Highway 6 about one mile north of the entrance to Waterton Park. Ponds are kettles.*

### *Waterton Townsite – Cameron Lake*

The road follows the lower slopes of Mount Crandell up the valley of Cameron Creek past exposures of Altyn-Waterton Limestone, the layers beneath the road being tilted down toward the valley floor. Apparently they are sliding because it has been necessary to patch the pavement in numerous places. Had the road been built on the opposite side of the valley where the strata dip into the hill, there would have been no problem with sliding.

A conspicuous monument on the south side of the road marks the site of the first oil discovery in western Canada. A well drilled here in 1902, to a depth of 1024 feet, produced small quantities of oil that generated major excitement leading in a few years to discovery of large oil fields on the plains of southern Alberta. The well is drilled into Precambrian sedimentary rocks but the oil must have come from Cretaceous sandstones beneath the Lewis Overthrust because Precambrian rocks almost never contain oil.

Between the oil discovery site and Cameron Lake, the road passes outcrops of green and red mudstone belonging to the Appekunny and Grinnell formations, and gray Siyeh limestone. Exposures of Siyeh Limestone along this road are the most accessible place in the Canadian portion of the Park to see a good variety of different kinds of fossil algae.

### *Waterton Townsite – Red Rock Canyon*

The road crosses extensive deposits of hummocky debris left by the glaciers that once filled the Waterton and Blakiston Valleys. A medial moraine marking the line where these two glaciers merged forms the pronounced grassy ridge at the upper end of the golf course.

Interesting structures are visible when the light is good on the face of Mount Crandell west of Blakiston Brook where it exits from the mountains. A large fault runs diagonally across the mountain bringing greenish-gray Appekunny mudstones over a streak of red Grinnell mudstones, a reversal of their usual order. Several very large and contorted folds are visible on the east and north sides of the same mountain which appears to have taken quite a beating, probably because it was near the leading edge of the overthrust sheet.

*Bertha Lake is a hanging valley scouped out by a small glacier tributary to the large Waterton Glacier. Canada Parks photo.*

*The pattern of braided channels in the gravels of Blakiston Brook is typical of streams heavily loaded with sediment.*

The short loop trail in Red Rock Canyon crosses many beautifully-preserved surfaces of suncracked and ripple marked muds laid down here over a billion years ago. The canyon gets its name from the red mudstones of the Grinnell Formation which outcrop in its walls. Boulders of dark gray limestone, which weather nearly white on old surfaces, have tumbled down from cliffs in the Siyeh Limestone. Many of these contain excellent specimens of fossil blue-green algae including stromatolites that look almost like heads of cabbage in the rock.

Downstream from the road, the trail crosses abruptly onto green mudstones of the Appekunny Formation. The last stop is at Blakiston Falls, about one-half mile beyond the bridge. The lip of the falls is in the trough of a sharp downfold, or syncline, in the sedimentary layers. Falls are often located in such places, presumably because the rocks have been tightly squeezed in the axis of the fold making them more resistant to erosion.

*Red Rock Canyon, a small gorge tributary to Blakiston Brook, has cut into hard Precambrian rocks since the end of the last ice age. Canada Parks photo.*

*Citadel Peaks jut into the skyline west of Waterton Lake. Mount Cleveland, at 10,438 feet, the highest peak in the Park, is in the distance on the east side of the Waterton Valley.*

## WATERTON LAKE BOAT TRIP

### *Waterton – Goathaunt*

Waterton Lake fills a long trough deeply-scoured into the hard Precambrian sedimentary rocks by a large glacier that had its main source in the central part of Glacier Park. Altyn-Waterton Limestone was especially resistant to glacial erosion so the ice rode up over it sculpturing the beveled ridge under the Prince of Wales Hotel and covering it with glacial debris. A corresponding ridge is on the opposite side of the lake.

The Waterton Valley was gouged by a very powerful glacier so the tributary streams now enter from hanging valleys, their mouths undercut by the large glacier in the main valley. Each of these tributaries has its source in a glacially-carved cirque.

Outcrops of red mudstones in the Grinnell Formation make a pronounced belt of color about halfway up the mountainsides on the west side of the Lake. Gray cliffs cut into Siyeh Limestone rise above the red band. Massive Goathaunt Mountain, east of the south end of the Lake, is carved mostly from greenish-gray Appekunny mudstones and capped by red Grinnell mudstones.

*Bedding surface in red mudstone of the Kintla Formation marked by impressions of cubical salt crystals. From an outcrop beside U.S. 2 a few miles north of Walton Ranger Station. Approximately natural size.*

Altyn-Waterton Limestone outcrops on the east side of the lake a short distance north of the international boundary forming Blacktail Point and the nearby cliffs. A very pretty series of small folds, each about 5 feet high, is exposed at the base of the cliffs immediately north of Blacktail Point, just above water level.

*West Glacier – East Glacier*

U.S. 2 skirts the southern edge of Glacier National Park for 56 miles between West Glacier (Belton) and East Glacier, a surprisingly easy and open route over the Continental Divide.

Between West Glacier and the mouth of Harrison Creek, about 7 miles to the east, U.S. 2 follows the Middle Fork of the Flathead River through a narrow canyon separating the Flathead Range to the south from the Belton Hills to the north. Long sections of the road are new, having been rebuilt after the disastrous flood of June, 1964. Bedrock is mostly mudstones belonging to the formations above the Siyeh Limestone which outcrop in the high peaks of the central part of Waterton-Glacier Park. Watch for salt casts in red ripple-marked mudstones. Here, west of the North Fork-Middle Fork Valley, these formations are at a much lower elevation.

At Harrison Creek the valley abruptly widens and the highway begins to follow a wide open floodplain which continues upstream for a little more than 20 miles following a remarkably straight northwest-southeast course. Straight river valleys are unusual and demand an explanation. In this case, straightness is due to the fact that the stream has eroded its valley along the line of the fault that defines the western boundary of the North Fork-Middle Fork Falley. Crushed and broken rock in the fault zone made it a line of weakness which the river exploited.

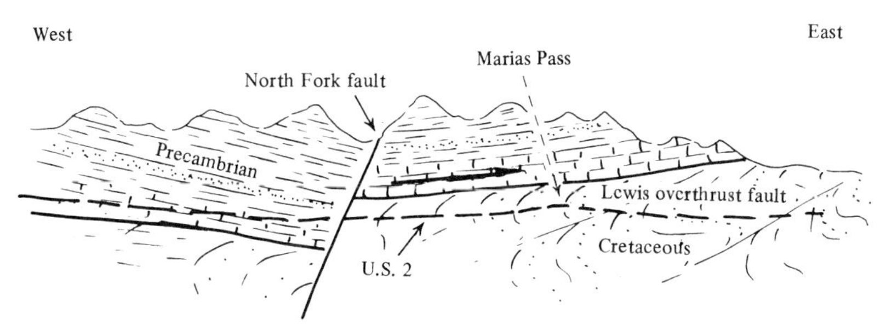


*Cross-section along the line of U.S. 2 from the mouth of Bear Creek, at Nimrod, to East Glacier. Area west of North Fork fault is the southern end of the North Fork Valley, here free of valley fill deposits.*

Between Nimrod and Marias Pass, U.S. 2 follows the valley of Bear Creek which pursues a very straight southwest-northeast trending course. The western half of this part of the route, between Nimrod and Giefer Creek, cuts across the southern end of the North Fork-Middle Fork Valley where it is not deep and contains no valley-fill sediments.

Between Marias Pass and Giefer Creek, U.S. 2 follows the southeastern edge of the slab of Precambrian sedimentary rock that moved eastward on the Lewis Overthrust Fault. Rocks north of the highway are part of that slab. Those south of the highway, although they belong to the same formations, are not part of the same slab. Instead, they belong to a series of westward-sloping slabs stacked on one another like fallen dominoes to make the Sawtooth Range. For several miles west of Marias Pass, Bear Creek has cut all the way through the Precambrian rocks exposing folded black Cretaceous mudstones beneath the Lewis Overthrust Fault.

The view north from Marias Pass is a geologic spectacular, several miles of continuous exposure of the Lewis Overthrust separating Precambrian rocks above from Cretaceous rocks below. The Precambrian rocks had to move many miles eastward, perhaps as many as 35, to get into this position above the rocks of the plains. No wonder that the first geologists to recognize this situation had trouble convincing their colleagues that they hadn't simply been lost in the woods.

*Panorama of view looking north from Marias Pass. The nearly-straight line of smooth talus slopes slanting gently upwards to the east (right) marks the Lewis Overthrust. Rocks forming slopes below are Cretaceous sandstones and mudstones about 70 million years old; those above are Precambrian limestones and mudstones more than one billion years old.*

Between Marias Pass and East Glacier, U.S. 2 follows the valley of Summit Creek and Railroad Creek, which have been eroded into folded Cretaceous sandstones and mudstones below the Lewis Overthrust. Good exposures are difficult to see from the road because these rocks are not resistant to erosion and in most places are covered by soil. But locally a few small folds are visible. Watch for outcrops of brown sandstone and black, shaly mudstone.

*The Middle Fork of the Flathead River has entrenched its valley into thick deposits of alluvial gravels. These were laid down during a time when the river transported a much heavier sediment load than it now carries.*

## U.S. 89 – THE BLACKFOOT HIGHWAY

### *East Glacier – St. Mary*

Between East Glacier and St. Mary, U.S. 89 winds its way for nearly 32 miles over a series of ridges, all remnants of the old pre-glacial high plains surface now deeply dissected by valleys eroded by streams and glaciers during the last 3 million years of the Pleistocene Epoch. Sweeping perspectives of the Park to the west and the high plains to the east open from the ridges above the valley floors which seem enclosed in a stunted forest of small aspens and gnarled pines. Most of the landscape is covered by glacial till but in places there are outcrops of greenish brown sandstone and black mudstone dating from the Cretaceous Period more than 70 million years ago.

Between East Glacier and the junction with the Two Medicine Road, about 4 miles north, the road stays in the valley of Two Medicine Creek. Forests growing near the road, and the low perspective from the valley floor, make it very difficult to get an impression of the geology.

Between the junction with the Two Medicine Road and Kiowa Junction, 12 miles north of East Glacier, the highway winds over the top of Two Medicine Ridge providing excellent views of the east front of Glacier Park and up the Two Medicine Valley.

The crests of the high ridges between East Glacier and St. Mary are capped by deposits of glacial till much older than any known elsewhere in the area. Prolonged weathering has given this till a distinctly reddish tint, contrasting to the brownish gray of most younger till in the area, and dissolved out most of the limestone boulders. Older glacial till is rare and difficult to find in the Rocky Mountains because in most places it has either been removed by erosion or buried beneath deposits of the last glacial period.

Evidently the older till was deposited by a vast sheet of ice that formed at the base of the mountains, fed by glaciers emerging from the mountain valleys, and then spread some miles eastward over the plains. Presence of older till high on ridge crests between East Glacier and St. Mary demonstrates that ice stood much deeper during at least one of the earlier glacial periods than during the one most recently ended.

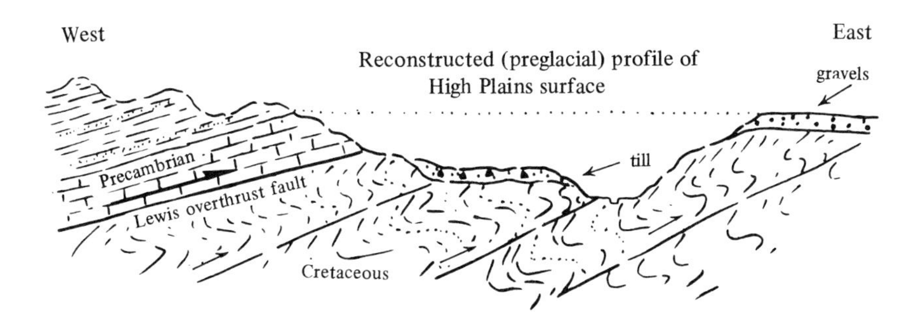


*West-east cross-section across the line of U.S. 89 between East Glacier and St. Mary.*

Younger glacial deposits are exposed on the lower slopes of Two Medicine Ridge. One roadcut on the south side of the ridge is especially interesting because it consists of beds of glacial outwash standing on end. Details of bedding in the layers of sand and gravel are hardly disturbed even though the material is quite soft. It must have been solidly frozen when an advance of the ice bulldozed it into its present position.

A roadcut on the north side of Two Medicine Ridge contains beautiful folds and small faults in layers of brown Cretaceous sandstone. This is the kind of structure that would be seen in the bedrock all along the eastern front of Waterton-Glacier Park if only there were more outcrops. Perhaps these rocks were rumpled beneath the Lewis Overthrust and are exposed now that their burden of Precambrian rock has been removed by erosion.

*Folded beds of Cretaceous sandstone on the north side of Two Medicine Ridge are cut by small faults at both ends of the picture.*

Between Kiowa Junction and St. Mary, U.S. 89 crosses the Milk River Ridge and north of it the St. Mary Ridge both capped by deposits of older glacial till. The Hudson Bay Continental Divide follows St. Mary Ridge. Cut Bank Creek and Fox Creek, both tributaries of the Milk River were named by Lewis and Clark who were impressed by the milky color of its water where it meets the Missouri River. Rock flour released by melting glacial ice in the headwaters in Glacier Park was responsible.

*Layers of sand and gravel glacial outwash standing on end in a roadcut on Two Medicine Ridge.*

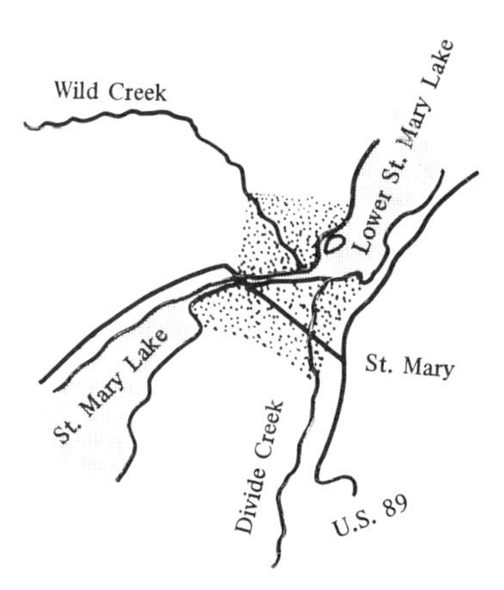


*St. Mary Lake is impounded by coalescing fan-shaped deposits of debris swept from Divide and Wild Creeks.*

Divide Creek, which U.S. 89 follows from the crest of the Hudson Bay Divide into St. Mary, was the source of much of the sediment that formed a large alluvial deposit in the St. Mary valley impounding St. Mary Lake. The rest of the dam is made of a similar deposit dumped from Wild Creek on the opposite side of the valley. No doubt these deposits were laid down within a relatively few years while the glaciers were rapidly melting.

*St. Mary — Chief Mountain Junction*

U.S. 89 follows the east side of Lower St. Mary Lake almost the whole 8 miles from St. Mary to Babb. Like its upstream sister, Lower St. Mary Lake is impounded by a dam composed of alluvium swept into the St. Mary Valley from a tributary, in this case Swiftcurrent Creek. The road crosses that alluvial deposit in the short stretch between the lower end of the lake and Babb.

The road beside Lower St. Mary Lake is built on glacial till belonging to a huge lateral moraine that lines the east side of the valley. Numerous roadcuts provide excellent exposures of glacial till, several are sliding because they were left too steep to be stable. Directly across the valley from Babb several large landslides mark places where the St. Mary River has eroded steep banks into the same moraine.

Between Babb and Chief Mountain Junction, the road follows the west side of the St. Mary River valley. The glacier that once filled this valley extended several miles farther before finally meeting the continental ice moving southwestward out of central Canada.

# Glossary

Most technical vocabulary is not really essential and very little is used in this book. The following glossary includes most of the geologic terminology likely to be encountered during a visit to Waterton-Glacier Park.

**Anticline.** A fold that bends rock layers upwards into an arch.

**Arete.** A knife-edged ridge gouged by glaciers on both sides. See photographs, pages 35, 36, 40 and 63.

**Argillite.** Very hard mudstone. Precambrian mudstones in Waterton-Glacier Park are hard enough to be called argillites, the Cretaceous mudstones are too soft.

**Basalt.** A black volcanic rock composed mostly of microscopic crystals of plagioclase and pyroxene. The Purcell flows are basalt and so are the fine-grained parts of the Purcell sill.

**Bedding.** Layering in sedimentary rocks. See photographs on pages 18 and 98.

**Bedrock.** Solid rock that has not been moved from its original position by any process of erosion. Includes the Precambrian sedimentary rocks and the Cretaceous mudstones but not glacial till or stream gravels.

**Bergschrund.** The gap that opens where the head of a glacier pulls away from its rock headwall as the ice moved downslope.

**Breccia.** A rock made of angular fragments cemented together. See photograph on page 71.

**Cirque.** A bowl-shaped hollow scooped out of the side of a mountain at the head of a glacier. See photographs, on pages 32, 35, 72 and 89.

**Col.** A notch cut in a divide where glaciers have bitten into it from both sides. Logan Pass is an example. See photographs, pages 35 and 36.

**Collenia.** One of the common varieties of stromatolite.

**Conophyton.** One of the less-common varieties of stromatolite.

**Diabase.** A dark-colored igneous rock composed of visible crystals of white plagioclase set in a matrix of black pyroxene. The coarse-grained parts of the Purcell sill are diabase.

**Dike.** A steeply-inclined sheet of igneous rock formed when molten magma injected across the beds in sedimentary rocks. See photograph, page 68 and diagram on page 14.

**Fault.** A fracture in the rocks, the opposite sides of which have moved.

**Firn line.** The lower limit of the area on a glacier in which the previous winters snowfall survives the next summer. See photograph, page 43.

**Formation.** Any body of rock that can be identified and traced over a considerable distance.

**Gletschermilch.** Water that looks milky because it contains glacial rock flour.

**Glacial drift.** All kinds of glacial deposits, both till and outwash, collectively.

**Glacial rock flour.** Rock dust and silt pulverized by the grinding action of a glacier.

**Granite.** An igneous rock composed mostly of visible crystals of quartz and feldspar. There are no outcrops of granite in Waterton-Glacier Park.

**Hanging valley.** A tributary valley left high by glacial deepening of the main valley. See photographs, pages 34, 35, 87 and 89.

**Horn peak.** A mountain that has been carved away by glaciers to a pointed shape. Many of the prominent peaks in Waterton-Glacier Park are horns. See photographs, pages 32, 36, 72, and 87.

**Kame.** A small hill of water-sorted and layered glacial debris in a moraine of unsorted till. Most kames consist of debris that washed into a hole or crack in the ice.

**Kettle.** A small pond formed where a block of ice buried by glacial deposits melts leaving a hole. The plains east of Waterton-Glacier Park are liberally dotted with kettles. See photographs, pages 38 and 88.

**Klippe.** A piece of an overthrust slab isolated by erosion. Chief Mountain, a mass of old Precambrian rock standing on much younger Cretaceous shale is a classic example. See photographs, pages 83 and 84.

**Limestone.** A sedimentary rock composed of calcite, the mineral form of calcium carbonate. The term limestone is used somewhat loosely in this book to cover both true limestone and dolomite, a very similar rock that includes considerable magnesium in its chemical composition.

**Marble.** Recrystallized limestone, consists of coarsely-granular calcite.

**Metagabbro.** Gabbro, an igneous rock similar to diabase, that has been recrystallized since the original magma solidified. The name has been mistakenly applied to the Purcell sill and dikes in Waterton-Glacier Park. Even though parts of these might reasonably be called gabbro, instead of diabase, they are not sufficiently recrystallized to merit application of the term "metagabbro."

**Molar tooth structure.** Fossil algae preserved as crumpled structures that sometimes suggest molar teeth. See photograph, page 7.

**Moraine.** A ridge of glacial till dumped around the margin of a glacier. See photographs, pages 36, 37, 81 and 88.

**Outcrop.** Any exposure of bedrock.

**Outwash plain.** A deposit of stream gravels forming a flat surface extending away from the mouth of a valley or the former front of a glacier. See photograph, page 38.

**Overthrust.** A type of fault in which an extensive slab of rock is moved across a nearly-horizontal surface. See photographs, pages 94 and 95.

**Pothole.** A cylindrical hole drilled into bedrock by pebbles turning on the stream bottom beneath a whirlpool. See photographs, pages 56 and 57.

**Recessional moraine.** A moraine similar to a terminal moraine marking a place where the ice paused short of its maximum advance.

**Scree.** A talus slope. A sliding slope of loose blocks of rock pried from a cliff by frost action. See photographs, pages 63 and 95.

**Sill.** A layer of igneous rock injected as a molten magma between beds of sedimentary rock. See photographs, pages 15 and 72.

**Strata.** Beds or layers of sedimentary rock.

**Striations.** Scratches rasped onto a rock surface by glacial scouring. See photographs, pages 30 and 45.

**Stromatolite.** Algal heads preserved in rock as stacks of nested shells resembling cabbages. See photographs, pages, 6, 11, 66 and 82.

**Syncline.** A downfold in the rock layers.

**Talus.** Blocks of rock pried loose by frost wedging. Usually forms sliding debris slopes. See photographs, pages 62, 63 and 95.

**Tarn.** An alpine lake occupying a basin hollowed out of solid bedrock by glacial erosion. A cirque lake. See photographs, pages 35 and 89.

**Thrust fault.** A fault in which older rocks are pushed up over younger ones along an inclined fracture surface.

**Till.** A glacial deposit composed of all sizes of material mixed together without layering or sorting. Till is dumped directly from melting ice. See photographs, pages 37 and 46.

# References

Despite its great interest, relatively little has been written about the geology of Waterton-Glacier Park. Most of what has been written is published in scientific journals not readily accessible except in large research libraries. Following is a short list of representative publications that are available in most univeristy and large public libraries. People seriously interested in reading further will find more comprehensive lists of references in some of the more recent of the works cited below.


Alden, William C., 1932. Physiography and glacial geology of eastern Montana and adjacent areas. U.S. Geological Survey Professional Paper 174, p. 133.

Daly, R.A., 1912. Geology of the North American Cordillera at the forth-ninth parallel. Geological Survey of Canada, Memoir 38.

Douglas, R.J.W., 1952. Waterton, Alberta: Geological Survey of Canada Preliminary Map 52-10.

Fenton, C.L. and Fenton, M.A., 1937. Belt series of the north; Stratigraphy, sedimentation, paleontology. Geological Society of America Bulletin, v. 48, pp. 1873-1969.

Obradovich, J.D. and Peterman, F.E., 1968. Geochronology of the Belt Series, Montana. Canadian Journal Earth Sciences, v. 5, pp. 737-747.

Rezak, Richard, 1957. Stromatolites of the Belt series in Glacier National Park and vicinity, Montana. U.S. Geological Survey Professional Paper 294-D.

Richmond, G.R. et al, 1965. The Cordilleran ice sheet of the northern Rocky Mountains, and related Quaternary history of the Columbia Plateau, *in* the Quaternary of the United States. H.E. Wright, Jr. and D.G. Frey, editors. Princeton University Press.

Ross, Clyde P., 1959. Geology of Glacier National Park and the Flathead region, northwestern Montana. U.S. Geological Survey Professional Paper 296.

Willis, Bailey, 1902. Stratigraphy and structure, Lewis and Livingston Ranges. Geological Society of America Bulletin, v. 13, pp. 305-352.